Project Management
Skills for Healthcare

Project Management Skills for Healthcare

Methods and Techniques for Diverse Skillsets

Lisa Anne Bove, DNP, RN-BC
Susan M. Houston, RN-BC, PMP, CPHIMS, FHIMSS

Routledge
Taylor & Francis Group

A PRODUCTIVITY PRESS BOOK

First published 2020
by Routledge
52 Vanderbilt Avenue, New York, NY 10017
and by Routledge
2 Park Square, Milton Park, Abingdon, Oxon, OX14 4RN

Routledge is an imprint of the Taylor & Francis Group, an informa business
© 2020 Taylor & Francis

Cover Art by Nix Houston

Trademark Notice: Product or corporate names may be trademarks or registered trademarks, and are used only for identification and explanation without intent to infringe.

Library of Congress Cataloging-in-Publication Data

Names: Bove, Lisa Anne, author. | Houston, Susan M. (Susan Murphy), author.
Title: Project management skills for healthcare : methods and techniques for diverse skillsets / Lisa Anne Bove and Susan M. Houston.
Description: New York, NY ; Abingdon, Oxon : Routledge, 2020. | Includes bibliographical references and index.
Identifiers: LCCN 2020007503 (print) | LCCN 2020007504 (ebook) | ISBN 9780367376499 (paperback) | ISBN 9780367403973 (hardback) | ISBN 9780429355882 (ebook)
Subjects: LCSH: Health services administration. | Project management.
Classification: LCC RA971 .B8355 2020 (print) | LCC RA971 (ebook) | DDC 362.1–dc23
LC record available at https://lccn.loc.gov/2020007503LC ebook record available at https://lccn.loc.gov/2020007504

Visit the Taylor & Francis Web site at
http://www.taylorandfrancis.com

and the CRC Press Web site at
http://www.crcpress.com

A catalog record for this title has been requested

ISBN: 978-0-367-40397-3 (hbk)
ISBN: 978-0-367-37649-9 (pbk)
ISBN: 978-0-429-35588-2 (ebk)

**Typeset in ITC Garamond Std
by Cenveo® Publisher Services**

Contents

List of Figures and Tables

Figures

Tables

Acknowledgments

I would like to thank my family and friends (too many to list individually) who have consistently supported my efforts. You are honest with me, and often help to redirect my personal and professional efforts and bring me back on track. I could not have accomplished this without all of you!

—Lisa Anne Bove

I would like to thank my wonderful family for their unwavering support. For my husband, Gary, you are the rock that keeps me focused and motivated. You are always there with words of wisdom or sarcasm, whichever I need at the moment. For my children, Nicole, Nick, Matt, and Dana, you have always been there whenever I have needed encouragement or graphics. Thank you.

—Susan M. Houston

We would both like to thank Nix Houston for the wonderful cover art.

About the Authors

Dr. Lisa Anne Bove is an Assistant Professor at the University of North Carolina Wilmington. Dr. Bove has worked in healthcare informatics for over 25 years in a variety of positions and has been certified by the American Nurses Credentialing Center in Nursing Informatics since 1996. Her teaching is focused on informatics, project management, and leadership. Her field of study focuses on implementation science and technology adoption in an effort to help improve nurse efficiency and advance practice through data. Dr. Bove actively advances the field of nursing informatics through involvement in the Healthcare Information and Management Systems Society (HIMSS), the American Nursing Informatics Association (ANIA), and local informatics nursing groups. She has published and spoken on numerous topics, including informatics and project management.

Susan M. Houston is a senior consultant in healthcare IT after retiring as the Chief of the Portfolio Office within the Department of Clinical Research Informatics at the National Institutes of Health Clinical Center. Her background includes clinical nursing, informatics, and project, program and portfolio management. Ms. Houston has presented at the local, regional, national, and international levels. She has authored a variety of articles and books on project management, application management and informatics. She is a member of the Project Management Institute (PMI), American Nursing Informatics Association (ANIA), and the Healthcare Information and Management Systems Society (HIMSS), while serving on various committees.

Chapter 1

Introduction

Healthcare organizations are always working to be more efficient while providing the best care possible. More evidence-based practice is being defined through the use of big data, analytics, and outcome measurements. Software and hardware are constantly being implemented and optimized to improve processes, documentation, and the bottom line. This focus on efficiency and outcomes leads many organizations to add multiple projects to their day-to-day management of patients. Because of this, many people who are assigned to the project have little experience with project management skills or the project management process, which can lead to project failure.

Project Definition

A project differs from day-to-day management of the organization. A project is defined as a "temporary endeavor undertaken to create a unique product, service, or result," and has a defined beginning and end (PMI, 2017, 4). While projects themselves are temporary, their results or deliverables exist beyond the end of the project. Projects drive change in organizations and can add to business value through increased market share, better outcomes, and/or better efficiency. Projects come in all sizes. They can be as big and complicated as building a new hospital or implementing a new electronic medical record across multiple facilities. They can also be smaller, but no less important or complicated, such as implementing a new evidenced based practice or teaching program.

Project Failure Rates

Whether big or small, however, many projects fail. And projects fail at an amazing rate. The Standish Group published their ground-breaking Chaos report on project statistics in 1994. In their initial study of 365 companies' information technology (IT) projects, they found 31.1% of projects would be cancelled before they got completed, 52.7% of projects would cost 189% of their original estimates, and only 16.2% would be completed on-time and on budget (Standish, 1994). Of the projects completed, only 42% of the originally proposed features and functions would be available. In addition, they found that larger companies had a lower success rate with projects – that is projects completed on time, on budget and on target – than small- or medium-size companies. In their 2015 report, they revised the definition of success to include six individual attributes of success, adding meeting goals, finding value, and satisfaction to the original attributes (Standish, 2015). Using the original attributes, 36% of projects were successful, while only 29% were considered successful with the revised attributes (Standish, 2015). One of the biggest changes in their 2015 study showed that smaller projects failed more than larger projects, 61% and 11% respectively.

While The Standish Group initially looked at IT projects, the success or failure of other projects is similar. Starting with their 2015 report, The Standish Group studied many types of projects, including projects in the banking, retail, government, and healthcare industries. They found that healthcare projects were only successful 29% of the time. In addition, they only looked at projects that were defined as projects by the facility studied. Many healthcare projects are not managed like projects but instead are something that a leader attempts to achieve and are probably not included in these statistics, which means failure rates in healthcare are probably much higher.

The Standish Group is not the only organization assessing project failure and the impact of these failures. A Harvard Business Review reported that over half of projects fail (HBR, 2012). In addition to lack of project management, they suggest project failure is due to the fact that people do not speak up about issues with the project. Many leaders assume they know how to implement a project and what the potential risks are, but often staff and even patients have valid concerns that need to be addressed. Both HBR and The Standish Group, as well as many other experts, have shown that project management can increase the probability of successfully completing a project on time, on budget, and with the expected scope and quality. In the 1994 Chaos report, the three major reasons for project success, in addition to project management, included user involvement, executive management support, and a clear statement of requirements (Standish, 1994). As projects and project management have evolved, successful projects should now also invest in smart, trained people, executive sponsorship, organizational emotional maturity, user involvement, and optimization in order to increase the likelihood of success (Standish, 2015). While many healthcare leaders are 'smart, trained people', project management skills are not usually part of their knowledge base.

Using the Project Management Process to Improve Success

In order to increase the likelihood of project success, each project should follow a standard project management process. Project management is defined as "the application of knowledge, skills, tools, and techniques to project activities to meet the project requirements and to obtain benefits and control not available by managing program components individually" (PMI, 2017, 10). Project management can help organizations to implement projects effectively and efficiently. In addition, project managers (PMs) can help increase the chance of project success by using standard tools and balancing the influence of constraints on the overall project.

Project management processes fit in the project lifespan and are discussed in five groups: Initiation, Planning, Execution, Monitoring and Control, and Closing process groups (PMI, 2017). Successful projects use tools within each of these project groups. For example, a business case defines why a project needs to be done. A project charter describes what needs to be done as well as what is required to complete the project, and a Work Breakdown Structure (WBS) or workplan describes when project tasks will be done. A communication plan describes how information during the project will be shared and a risk management plan helps to plan for potential issues during the project. Project managers (PMs) use these tools throughout the project to manage tasks, scope, and resolve issues.

Purpose and Chapter Overviews

The purpose of this book is to introduce the project management processes to novice project managers and discuss project management tools used at any level of experience. This book is intended to be used by IT and non-IT project managers. Chapter 2 (Project Management Processes) describe the overall standards and knowledge domains of project management. This will set the stage for the rest of the chapters. In addition, best practices in each of the process groups will be discussed. Chapter 3 (Novice to Expert) will address Dreyfus' learning continuum from novice to expert. In this chapter, we will discuss how this model describes how knowledge acquisition occurs and provides a method to assess and support the development of skills and competencies. PMs can move through the five stages – novice, advanced beginner, competent, proficient, and expert – as they develop project management skills. Expert PMs do not stop learning; rather, they continue to evaluate their practice and learn new skills to keep up-to-date. Throughout this book, we will use the term *experienced*, rather than *expert* to support this tenant. Each of the following chapters will focus on the project management process groups. Each process group will be discussed first at a novice level and then at a more experienced PM level. Tools that can be used at each level will be addressed as well.

Chapter 4 (Initiation – Novice) will discuss the initiation process group. Initiation begins with the request for a new project and ends when a decision is made related to the authorization of the project. Tools discussed in this chapter include the request for the project and a project charter or business case. The terms *stakeholders* and *sponsor(s)* will also be described. Chapter 5 (Planning – Novice) begins as the project authorized and resources are available, after the initiation phase is completed. The planning process group includes the processes and tools that are needed to successful determine how a project will be managed. In this chapter, we will discuss how the PM will focus on creating the project management plan that includes the workplan with assumptions and constraints and finalize scope and communication plan. As discussed in the chapter, without proper planning, the project is less likely to be successful.

Once planning is complete, the PM will begin to focus on the actual work of the project. Chapter 6 (Execution, Monitoring and Control – Novice) will address how a novice PM will carry out the project management plan and performing the scheduled activities. Along with the execution process group, this chapter will discuss the simultaneous monitor and control process group. These process groups are carried out together until the project reaches the closure phase. The execution phase is where the work of the project gets done. It includes 'build', test, training, and go-live activities. The monitor and control phase is simultaneous with the execution phase and includes the main work of the PM during these phases. Tools discussed in this chapter include managing scope and requirements, status reporting and communication, and issues and risks management. Chapter 7 (Closing – Novice) will discuss how the project is finalized. The closing process group begins when the project is live, but doesn't end until the project deliverables are accepted by the sponsor(s), documentation is completed and archived, and resources are released. In addition, the PM will complete a completion document which will describe the project metrics and outcomes.

Chapter 8 (Initiation – Expert) will begin the second half of the book. In this, and the following chapters, the project management process groups will be discussed at a more complex level. Up to this point in the book, basic tools that the PM would use in each phase of the project were discussed. In the expert chapters, additional tools will be discussed that will help a PM manage larger, more complex projects. Chapter 8 (Initiation – Expert) will discuss creating a more detailed project charter, as well as the process to submit a project request. Chapter 9 (Planning – Expert) expands on the previous planning chapter to add more tools the PM can use during the planning phase. More details about how to create the workplan and detailed change request, issues, risks and quality management plans will also be included. In addition, organization success factors will be addressed. Chapter 10 (Execution, Monitoring and Control – Expert) will describe tools that will help the PM to execute larger, more complex projects. In this chapter, we will also discuss additional constraints that the PM needs to manage in order to move the project successfully to the closing phase. The last chapter, Chapter 11 (Closing – Expert) will discuss completing the project and closing out the metrics and other project deliverables. In addition, in this chapter, we will discuss transitioning the project to support.

Case Study

The case study described below includes projects that could be assigned to a novice or more experienced PM. Throughout this book this case study will be used to describe example tools PMs use to successfully implement projects. Different tools will use different projects within the case study to illuminate the way a project manager uses them throughout the project. We are using the project examples, not to create full project management plans, but rather using the various projects listed to give as many examples as we can.

According to *Health Facilities Management*'s 2017 Hospital Construction Survey, conducted in cooperation with the American Society for Healthcare Engineering (ASHE), 14% of hospitals surveyed have an acute care build project underway. Seven percent have a specially unit or building underway, and 6% are building a critical access facility. Ninety-five percent of respondents said they involve clinical and nonclinical staff in the design process of a new facility. They estimate it takes 7–8 years to design and build a major hospital, and during that time, clinical practice can change, so teams need to be responsive to change.

Your organization is building a new building and re-organizing patient care units. The Tower, as it is commonly called, will require new staff, new computers and communication devices, and patient equipment. In addition, policies, procedures, and electronic medical record documentation will be reviewed and potentially revised. While there is a construction team who is responsible for the actual building and supplies, your team needs to help develop and implement the plans to:

■ Create and deliver an orientation plan (training materials) for the new units
■ Locate hardware
■ Review policies and procedures around the admission, discharge, and transfer processes
■ Revise the electronic medical record to include the new units, locations, and service lines, as well as any required new documentation
■ Order, configure, test, and install the new computer (handhelds and carts) equipment Tower before patients are moved, and
■ Facilitate the patient move (go-live plan)

You will be working within the larger Tower project but will only be responsible for a specific project as listed above. You will be working with the IT, Education, Quality, and Facilities departments, as well as the construction team during the transition.

Chapter 2

Project Management Process

In this chapter, the project management process will be defined. The process groups and knowledge areas, defined by the Project Management Institute (PMI), will be described as well as how they work together during the project's lifecycle. The chapter will end with a description of how to tailor these best practices to meet an organization's need and specific culture. Before anyone can understand project management, it is important to first be familiar with the definition of a project. A project is temporary and produces a unique product, service, or result. It is progressively elaborated, or defined, as the project progressed through its lifecycle. There is a beginning and an end to each project. The beginning is when the need is identified, and the end is when the final defined objectives are met and are accepted by the sponsor(s). A project may also end if there is a decision to cancel if the objectives cannot be met, the need no longer exists, or the necessary recourses are no longer available.

The Project Team

As projects are temporary, project teams are also temporary. While there are organizations with teams that continuously work together one project after another, these are very rare. The more frequent situation has staff coming together for a single project based on the specific needs of the project. They work together only until the project is complete. It is also rare for the team members to be dedicated to only one project. They may be assigned to multiple projects and/or have other duties such as systems maintenance.

This provides some complexity to the project planning where the project manager (PM) has to ensure the tasks are scheduled for when the resources are available.

Project management is "the application of knowledge, skills, tools, and techniques to project activities to meet the project requirements" (PMI, 2017). Project management has been around since the mid-20th century, and it was introduced into the healthcare industry with the rise of the use of software applications. As healthcare organizations began implementing software, the vendors introduced the project management concepts. As the use of information technology (IT) expanded, organizations began hiring PMs and then began setting up project management offices (PMOs).

The Project Manager Role

The role of the PM is to facilitate and control the project as it moves through its lifecycle. There are many words that can be used to define the PM. They are organized and track all details while keeping an eye on the larger picture and how any issue or change would impact the project as a whole. They are motivators who keep the project team and the rest of the stakeholders moving forward and focused on the tasks and activities. Communication is one of the biggest roles of the PM as most, if not all, communication is sent or received by them. Communication occurs with the project team, sponsor(s), end users, and the larger stakeholder community. Information is shared verbally, through presentations, or written.

Based on the above roles, there are a wide variety of skills required for any PM to be successful. Below is a list of a few of these skills.

- Project Management Knowledge – having working knowledge of project management concepts, practices, tools, and vocabulary
- Business Domain Knowledge – having enough understanding of the business part of the project to be able to ask the right questions from subject matter experts (SMEs) and know when the answers are accurate. Understanding what is feasible, or not, or what is beneficial
- Communication – the ability to provide clear and concise communication to all stakeholders, at the appropriate level of detail, whether verbal, non-verbal, written, or visual
- Leadership – providing direction, encouragement, and guidance while maintaining control of, and accountability to, the project

- Negotiation – having the ability to lead a discussion while reaching a beneficial agreement for all involved
- Risk Management – the ability to identify, analyze, and control potential threats to the project
- Critical Thinking – having the ability to objectively analyze and evaluate an issue to form a judgment or decision
- Decision Making – being able to make a decision, when necessary, based on the information available
- Prioritization – the ability to identify what items are more important than others, such as conflicting tasks or activities
- Political and Cultural Awareness – the awareness of the political nature and culture of the organization and how they impact the acceptance of the project
- Organizational Skills – the ability to arrange and coordinate time, workload, resources, schedule, and prioritize projects
- Planning Skills – the activity of preparing and forecasting what is needed to complete a task or project to successfully meet the goals and objectives
- Proactive – the ability to control a situation through preparation or actually causing something to happen, rather than reacting afterwards
- Time Management – the ability to use time effectively and productively to ensure work is completed on or before the due date
- Cost Management – managing and controlling expenses to match the assigned budget
- Problem Solving – the ability to analyze issues or problems and find solutions
- Adaptability – being able to adjust to changes or new conditions
- Conflict Management – the process of limiting the negative aspects, while increasing the positive aspects, of conflict with the goal to enhance outcomes or performance
- Task Management – being able to manage a task, or activity, through its lifecycle, until completion
- Quality Management – the act of overseeing tasks or activities to ensure the output meets or exceeds expectations
- A Sense of Humor – used to help alleviate the stress and pressures of project management and project work by the project team
- Motivation, Team Building, Coaching, and Influencing – the ability to bring resources together and build a project team where they work cohesively and collaboratively focused on the goals and objectives of the project

Project Management Process Groups

There are five process groups as defined by the PMI: Initiation, Planning, Execution, Monitoring and Control, and Closing. While these are well known in the project management community, many projects are further broken down into different phases through the process of tailoring the basic framework. No matter how many phases, or their names, they all fit into the basic process groups. Tailoring will be discussed in further detail at the end of this chapter. The five process groups are discussed below.

The **Initiation** process group includes all activities leading up to formal authorization to begin the project. These activities may start or be completed prior to a PM being assigned and is dependent on the organization's methodology. These are important activities that provide an understanding of the business need and project objectives. All requests require some data gathering and analysis prior to the approval decision. In some organizations, this is more formal than in others, and the level of documentation will differ as well. It is best if all new requests are managed following a consistent process to ensure all are treated the same and produces a final deliverable that allows for a sound approval decision.

A new request is usually in response to some problem, an opportunity, or a new business requirement. The analysis of the request should include the evaluation of options for meeting the need and the relationship between the project and the organization's strategic plan. The main deliverable during this process group is the project charter, but a business case may be appropriate. The project charter defines the project to be completed, including the objectives, expected deliverables, estimated duration, and a forecast of the required resources. A list of typical contents of a project charter is below. The business case is used when the details of the project may be unknown, such as when the need is defined, but options for meeting it are still being evaluated. Each option may have different project requirements, making the completion of a charter difficult. If the business case is used, the authorization decision should include which option to implement.

The typical contents of a project charter include:

■ Project title
■ Estimated start and finish dates
■ Business need, goals, and scope

- Justification and background
- High-level deliverables
- Project manager and authorized level of authority
- Key project stakeholders
- High-level estimate for human resources
- Estimates for other resources
- High-level milestones and estimated timeline
- High-level budget estimation
- Strength, Weakness, Opportunity, and Threads (SWOT) analysis
- Assumptions and constraints
- Risks and risk management strategy
- Critical measures of success
- Review comments and approval decision

The initiation process group ends with the authorization of the project. While an individual with this authority can make the approval decision, it is more common to be made by a committee. A governance or steering committee may be responsible for reviewing and approving strategic plans, overseeing major initiatives, and/or allocating resources. Based on the committee's charter, it may also review and approve new IT requests while understanding the current workload and resource capacity. They would also make decisions on prioritization of work based on the limited resources and current organizational needs.

The **Planning** process group utilizes all the information gathered during the initiation process group to plan the completion of the project. In many organizations, the PM is not assigned until after the project is authorized, which gives the appearance that the initiation did not occur. Also, the project planning documentation rarely reflects the work completed during initiation. During planning, the PM expands on the work completed previously and creates the project management plan, which is actually a collection of plans defining how the project will be managed and completed. Some plans are drafted as part of the project management methodology and are used consistently across all projects, while others are unique to each project.

The plans that should be consistently used across all projects are related to how different portions of the project will be managed. These include management plans for Scope, Risks, Issues, Stakeholders, Costs, Requirements, and Changes. The plans define how each are identified,

documented, analyzed, and approved if necessary. They include roles
and responsibilities as well as any tools or templates that should be used
for each. They provide consistent approaches to the management of
each project to set expectations for responsibilities, communication, and
deliverables.

The plans that are unique to each project are often the deliverables from
the plans above. These documents include the following:

- Scope – sets the boundaries for the project by defining what is in and
  out of scope. This document usually includes the following:
  - Project name
  - Brief description
  - Scope statements
  - Justification
  - Project team members and their role
  - Assumptions and constraints
  - Timeline and major milestones
  - Deliverables
  - High-level budget
  - High-level implementation plan
  - Critical success factors
  - Approvals
- Budget – defines the budget for the project
- Stakeholder analysis – defines the stakeholders, their interest in the
  project and the details around their expected communication (i.e.,
  level of detail, frequency, format), and how they define success for
  the project
- Communication Plan – based off of the stakeholder analysis, this plan
  defines the who, what, when, how, and responsibilities of project
  communication
- Project Team's Roles and Responsibilities – outlines the roles and
  responsibilities for each member of the team. The role on the project
  may be different than their title within the organization, such as a
  department head who is the project sponsor
- Implementation Plan – defines how the project will be implemented.
  This could be in phases by functionality, in phases by department or
  patient care unit, as a big-bang with all functionality and all areas at
  once, or a pop-bang with a small pilot then everything and everyone
  else after

- Work Breakdown Schedule (WBS) – a hierarchical visual representation of the work to be done as broken down into smaller components. Visually, this may look similar to an organizational chart
- Workplan and Schedule – a list of tasks and deliverables along with their estimated start and end dates, work effort, and resources. This can be developed from the WBS or from historical details of previous projects that are similar. The high-level schedule can be represented by a calendar view of the work to be done for stakeholders who prefer a higher level of detail
- Quality Plan – defines how quality will be measured for the project, the roles and responsibilities for quality measurements as well as timeline and expected outcomes
- Metrics Plan – defines how success is defined for the project and may be combined with the quality plan. The plan should include metrics for a successful project, which are measured prior to the end of the project, as well as the details on how to measure, how well the business need was met, or if the expected benefits were seen. The latter is often measured outside of the project after 6, 12, or even 18 months, although baseline metrics may be collected during the project
- Risk Register or Log – the documentation of the risks for this project as well as the analysis and management strategies

By the end of the Planning process group, the project sponsor or sponsors should approve the project management plan, and a kick-off meeting is held. The kick-off meeting should include all project team members and key stakeholders. The agenda includes communicating the scope of the project, the implementation plan, and the project management process. The process discussion should include topics such as how changes to the project are reported and managed to avoid scope creep and how issues/risks are reported. The expected communication related to status updates is also an important component. This essentially kicks off the execution of the plans that have been defined and approved. In some organizations planning ends upon the sponsor(s)' approval, and the kick-off meeting occurs as the first activity of project execution. Since the activity marks the end of planning and beginning of execution, either location is acceptable.

The **Execution** process group is where all of the activities defined in the project management plan are completed. The PM facilitates the people and activities to ensure the completion of the work as scheduled. The project spends the most amount of time in this process group, along with monitoring and control, and the two occur concurrently. The PM must manage competing priorities especially when resources are not fully dedicated to the project work. Scope, time, cost, and quality need to be balanced to ensure a successful project as any change to one of these will impact one or all of the others.

The PM needs to work within the constraints of the assigned project team members and their availabilities to build a collaborative team. The team should provide the PM with regular updates as tasks are completed so the schedule can be kept current and for status updates to stakeholders per the communications plan. The PM is the primary point of contact for project information for both incoming and outgoing communication. All issues should be reported to the PM as soon as they arise so they can be evaluated and a resolution identified as quickly as possible. As work progresses, the project documentation is updated and shared. This includes the workplan where the PM keeps track of tasks that are completed, in progress, or next to start. Tasks may include development, documentation, testing or validation, end user training, and preparing for the transition to support.

The work to prepare and complete the activation, or go-live, is included in the execution process group. This activity follows the implementation plan and should include all work surrounding the final deliverable to the end users, whether in phases or all at once. There is a lot of variability of work to be completed during this activity, depending on the final deliverable(s) and what is needed to make it available to the end users. Proper planning for this activity is a must as first impressions are key.

The **Monitoring and Control** process group occurs concurrent to the execution process group, as mentioned above. The monitoring activities help to identify potential problems, and regular performance measurements identify variances to the project management plan. Controlling changes within the project is an ongoing effort for any PM. Each change should be evaluated for how it will impact the project as a whole and that they provide a benefit. Changes to the approved scope or requirements can have a big impact to any project. Careful analysis of any request, prior to the approval decision, helps to set expectations when the schedule, budget, or resources are impacted. The control of the project boundaries is the best defense against scope creep.

Periodic review of the project performance can ensure the quality of the project's deliverables. This can be done through a variety of tools depending on the unique project and the final product. Some of the tools available are quality audits, benchmarking, testing, or validation. The final deliverable(s) may be high quality, but if they do not match the requirements, or meet the need, they will not be beneficial to the users or the organization. The continuous monitoring and reevaluation of risks occur throughout the project. The risk response plan should be updated as new information is available, and they should be closed once the timeframe for the risk passes. If risk management is done correctly, it is often unnoticed since the responses are planned ahead of time and implemented quickly to resolve the problems as they arise.

The **Closing** process group includes key activities to bring the project to a successful end once the project's deliverables have all been provided and accepted by the project sponsor(s). During this time, the project team remains involved to assist end users until the project is officially transferred to the support staff. All project documentation should be finalized and archived for use as historical information to future projects. The final evaluation of the project success factors should be completed to demonstrate the effectiveness of the project. All stakeholders are brought together to discuss any lessons learned during the project as a retrospective review of the project activities. The lessons learned feed into improvements of processes and activities for future projects. It is important for all to understand this is a learning activity and not a time to point fingers and no comments or input is wrong.

The formal acceptance of the project's closure comes from the sponsor(s). The closure document is an evaluation of the original project scope and any approved changes and provides a gap analysis of what was planned vs. what was actually completed. The sponsor(s)' signature provides approval that all requirements were met, they accept the final product(s), and approve the closure of the project. The celebration is the final activity before the resources are released and the project is formally complete.

All projects move through these five process groups, but the amount of time spent in each one, and the specific activities, are dependent on the uniqueness of each project. Throughout the project's lifecycle, there are defined activities and tasks that fit into specific project management areas of knowledge and skills. The PMI has defined ten knowledge areas for effective project management. PMs should have knowledge and skills in each of these areas or have resources that can assist for those areas. Each of the five process groups defined above can be mapped to these knowledge areas. The PMI identifies the PM's activities for each knowledge area as listed below.

Project Management Knowledge Areas

Integration Management is used to coordinate all activities that occur through all process groups and knowledge areas. The main activities during integration management involve developing and managing the project management plan, which is a clear concise plan for how the project will be completed. The PM should focus on the big picture while ensuring the detailed tasks are completed on time. They must stay on top of the day-to-day activities while managing and controlling the big picture of the overall project. Activities for this knowledge area include:

Initiation Process Group
  – Develop project charter
Planning Process Group
  – Develop project management plan
Execution Process Group
  – Direct and manage project work
  – Manage project knowledge
Monitoring and Control Process Group
  – Monitor and control project work
  – Perform integrated change control
Closing Process Group
  – Close out project or phase

Scope Management starts with scope definition and management. These alone will not guarantee a successful project. Obtaining consensus of what is, and is not, included in the scope can be difficult based on the number of stakeholders involved. The scope defines the boundaries, what is and is not part of the project, how the project success will be defined, and feeds into the resources required and the schedule. The PM's role is to ensure the entire team understands the boundaries of the project once the scope is approved. Scope change control involves managing and controlling any changes to the project scope. Properly executed, this will avoid scope creep. Each change should be evaluated for the impact to the project as a whole and reviewed with the sponsor(s) for approval. This provides the sponsor(s) the ability to understand how each change will affect the cost, schedule, required resources, etc., for the project. Activities for this knowledge area include:

Planning Process Group
- – Plan scope management
- – Collect requirements
- – Define scope
- – Create WBS

Monitoring and Control Process Group
- – Validate scope
- – Control scope

Schedule Management involves all activities related to completing the project and is not as easy as it sounds. The PM works with the project team to identify the tasks necessary to complete the approved scope and deliverables. This includes the expected work effort, duration, resources, and cost to develop the schedule. Understanding the relationship between tasks ensures proper sequencing for the overall timeline. Once the schedule is defined and approved, it must be controlled and only adjusted with approved scope changes. Activities for this knowledge area include:

Planning Process Group
- – Plan schedule management
- – Define activities
- – Sequence activities
- – Estimate activity durations
- – Develop schedule

Monitoring and Control Process Group
- – Control schedule

Cost Management is related to managing the project budget. The total cost of the project is based on the scope, requirements and estimated activities. The budget may also include the cost of software, hardware, handheld or patient care devices or assistance from a vendor or consulting company. Budgets are often very tight with very little room for any adjustments, which is why the costs must be closely monitored to avoid unnecessary changes. In some organizations, the PM does not develop or manage the budget, but they should be aware of the associated costs and could be asked to provide input as line items are completed. Activities for this knowledge area include:

Planning Process Group
- Plan cost management
- Estimate costs
- Determine budget

Monitoring and Control Process Group
- Control costs

Quality Management has a variety of definitions, and each stakeholder may have their own way of defining quality for this project. Some options for quality are how the project requirements are met, how well the final product meets the indented use, or the detailed review of the completeness of each deliverable produced. The primary purpose of quality management is to ensure the final product meets the business needs. It is best to understand the way quality will be measured during project planning and include this in the project management plan. Activities for this knowledge area include:

Planning Process Group
- Plan quality management

Execution Process Group
- Manage quality

Monitoring and Control Process Group
- Control quality

Resource Management used to be called human resource management and is focused on determining what resources are needed, gathering the right team members and tracking their performance throughout the project. Depending on the resources assigned, the PM may need form them into a team that will work together to complete the project. The PM will need to negotiate for the necessary skills for the project team, ensuring the staff has available time to work on the project when they are needed in the schedule. It is rare for the members of the project team to report directly to the PM so they will need to work closely with their human resource manager to prioritize conflicts. In these situations, the PM has the authority to direct the staff's work within the project only. Activities for this knowledge area include:

Planning Process Group
- Plan resource management
- Estimate activity resources
- Acquire Resources

Execution Process Group
- Develop team
- Manage team

Monitoring and Control Process Group
- Control resources

Communication Management is more than just distributing information; it requires understanding of the information received. The communication plan defines what information to share with which stakeholders, the level of detail expected along with the frequency and mode. This plan includes what information the PM will send and receive from the stakeholders. The communication may be verbal, non-verbal, written, or visual. The PM spends most of their time communicating in one form or another. Activities for this knowledge area include:

Planning Process Group
- Plan communications management

Execution Process Group
- Manage communications

Monitoring and Control Process Group
- Monitor communications

Risk Management is the process of identifying, analyzing, and responding to risks within the project. This is one of the most underrated and forgotten parts of project management. The identification of risks is the responsibility of all project team members. Proper risk management is a form of insurance to lessen the impact of potential adverse events. Early identification of risks allows time to analyze and determine options for response. Most organizations struggle to balance risk and opportunity, and each has their set risk tolerance. Activities for this knowledge area include:

Planning Process Group
- Plan risk management
- Identify risks

    – Perform qualitative and quantitative risk analysis
    – Plan risk responses
Execution Process Group
    – Manage risk
    – Implement risk responses
Monitoring and Control Process Group
    – Monitor risks

Procurement Management is the process around obtaining and managing goods and services from outside sources. Other terms used are purchasing, contracting, or outsourcing. The PM should have a general understanding of the procurement processes to ensure they remain compliant with the regulations during communication and activities with these vendors. It is rare for the PM to actually manage the contracts; rather they work closely with the contracting department to ensure requirements and deliverables are defined while providing communication as each are met. In these instances, the contract staff is part of the project team. When contractors are brought in to supplement the project team, the PM should have the authority to direct their work and provide feedback on their performance. Activities for this knowledge area include:

Planning Process Group
    – Plan procurement management
Execution Process Group
    – Conduct, or assist with, procurements
Monitoring and Control Process Group
    – Control procurements

Stakeholder Management is an important activity for the PM. Everyone impacted by the project is a stakeholder, which includes members of senior leadership, department heads, end users, the project team, and anyone external to the organization that is impacted. The external stakeholders in healthcare projects could be patients, families, insurance companies, or community partners. Early identification of the stakeholders and the analysis of their expectations of the project are key to developing the communication plan as well as other portions of the project management plan. A stakeholder analysis is conducted to identify their interest in the project, how they measure the success of the project, what information they need along with frequency and mode of communication, how they define the boundaries of the project (what is in and out of scope), and how they can provide assistance to the PM and project team through the project. Activities for this knowledge area include:

Initiation Process Group
– Identify stakeholders
Planning Process Group
– Plan stakeholder engagement
Execution Process Group
– Manage stakeholder engagement
Monitoring and Control Process Group
– Monitor stakeholder engagement

Project Management Methodology

The information above, about process groups and knowledge areas, is a high-level overview of best practices. It is rare that any organization follows these as written without any modifications. Organizational culture, needs, project management skills, and types of projects all play a part in the project management methodology used. The concept of tailoring the best practices while retaining the key concepts is used across the project management industry and even in healthcare. An understanding of these best practices is a must before the process can be tailored to ensure the final methodology includes the basics. The following chapters will expand on what activities and deliverables are necessary, even for novice PMs, and which should be includes as the PM skills and methodology becomes more mature and why they are important.

The defined methodology should provide basic activities and deliverables for each phase of the project. Usually, this follows the five process groups and fits into a basic project lifecycle.

1. Project start (Initiation)
2. Project organization (Planning)
3. Completing the plan, doing the work (Execution, Monitoring and Control)
4. Project ending (Closing)

Some organizations do not involve the assigned PM until the project is approved, so the process may appear to start with planning. The best option in this case is to still define initiation, but identify who is completing the work during that phase. The methodology should grant the PM the authority to modify the process further depending on the type or complexity of the project. The required tasks and activities for a software application implementation are different than updating or documenting new policies and procedures. The quantity and type of deliverables may also be different.

The PM may choose to break the project down into additional phases, beyond the four mentioned above. For the software application, the phases could be planning, design, install, configure, test, train, go-live, and closing. The policy and procedure project may include phases for planning, gap analysis, documentation (draft, review, final, approval), effective date (go-live), and closing. The additional phases allow the PM to group the work to be completed into smaller buckets that may provide better understanding for stakeholders and ease of tracking. The actual amount or name of the phases for any individual project can vary based on need and what works best for the uniqueness of the project.

The very basics for any project should include the management of the scope, work, and communication. The management of the scope would include defining the project boundaries and obtaining approval. The approval step is as important as the boundaries as it puts a freeze on the scope. Any requested changes, once approved, should follow a standard process for analyzing the impact and obtaining approval where changes to the project are known so expectations are set if approved. This prevents scope creep, which is where changes are made unchecked and unapproved with the outcome being an out of control project with probable schedule delays and cost overruns. As the scope is controlled throughout the project, during closing, the PM should obtain official approval to close the project through the closure document.

The management of the work includes working with the team to create a workplan or schedule, which can be as simple as a spreadsheet or calendar to a formal WBS and workplan using project management software. How complex, the level of detail of each task, and the tool used, may depend on the complexity and duration of the project. Control of the work to be done is a must to ensure tasks are completed on time so the project stays on schedule. The control and management of risks and issues identifies when the schedule is impacted. The workplan or schedule is updated regularly as tasks are completed and modified if they are done early or delayed. Planning and executing the final deliverable, or go-live is also as important and the rest of the work in the schedule and is also based on the project's objectives. The final few tasks for the PM should include understanding the lessons learned and ensure all project documentation are complete, final, and archived.

The facilitation of all communication should follow the communication plan. The plan could be a detailed document outlining all communication based on the stakeholder analysis, or it could be a simple table with the 'who, what, when, how, and responsible party' for each communication channel. Following the plan to ensure all stakeholders have the information requested or needed for their responsibilities is an important role for the PM.

A PM methodology is a framework with practices, procedures, templates, and rules. The process should allow for tailoring for smaller, less complex, projects managed by a novice PM as well as a large, very complex, highly visible, project managed by an experienced PM. The methodology can be obtained from vendors, consultants, a government agency, or developed by those within the organization. No matter where it begins, it should be tailored to meet the needs and expectations of the organization where it will be used. The methodologies will differ between organizations, but the basics will always be present. The following chapters will further define these practices for the novice and advanced PM.

Chapter 3

Novice to Expert

In this chapter, we will discuss the novice-to-expert model originally identified by the Dreyfus brothers. This model describes how knowledge acquisition occurs, and provides a method to assess and support the development of skills and competencies. Project managers (PMs) often learn on the job when they are assigned to a project, rather than through formal education. Project management skills, however, do not come from just doing the job, but instead from progressing through various stages of expertise. The five stages – novice, advanced beginner, competent, proficient, and expert – are stages that we move through as we acquire skills. Expert PMs do not stop learning; rather they continue to evaluate their practice and learn new skills to keep up-to-date. Throughout this book, we will use the term *experienced* rather than *expert* to support this tenant. Understanding the difference between novice and experienced PM skills and tools will help the novice PM grow.

While project management has been around for centuries, few leaders in healthcare outside of information technology (IT) departments know the processes and tools needed to successfully plan and execute a project. Project management skills, however, can help to improve the likelihood of any project success from updating policies, changing staffing patterns, and implementing a new service line or evidence-based initiative as well as IT projects. Often, PMs are assigned to clinical or non-IT projects without formal training and are expected to follow standards and guidelines. Learning the skills needed to successfully manage a project can be done through both formal and informal education as long as the PM can focus on the skills needed.

Dreyfus Model

The Dreyfus Model of Skill Acquisition describes how learners acquire skills through instruction and practice. The Dreyfus brothers originally proposed this model in 1980 through their research at UC Berkeley and the United States Air Force. They postulated that students' progress through five stages: novice, advanced beginner, competent, proficient, and expert. Patricia Benner (1982) applied this model to nursing education and practice, and it has been used to understand how clinicians learn. The model has also been applied in a variety of professions, including engineering, telecommunications, financial services, and education. The Dreyfus model is used to provide a means of assessing and supporting the development of skills and competencies and to provide a definition of acceptable levels of capabilities in each stage.

Novice Learners

The first stage in the Dreyfus model is novice. According to this model, novices have little situational perception or discretionary judgement. They do not know how processes work in their new environment and cannot usually judge how events will proceed. They tend to rigidly adhere to rules and guidelines. Novices new to project management processes and tools need clear tasks and timelines. They often do not know enough to plan for risks and manage scope creep. Detailed project management plans and timelines, as well as clear prioritization, help project management novices to work through a project. PMs at this stage need clear direction from more experienced PMs. Novice PMs will often learn from their own mistakes rather than have experience from previous projects. They should only be assigned to small, less complex projects.

Advanced Beginner Learners

Advanced beginners can start to use guidelines along with their prior experience. Their situational perception is still limited, and they have minimal ability to prioritize tasks and issues. They have a working knowledge of the activities they should do but often just follow guidelines rather than understand them. Advanced beginners in project management can start to apply their previous project experience to the current project but may not be able to correctly adjust plans and/or plan for risks. Advanced beginner PMs start to function independently in smaller, less complex projects. They need guidance when projects hit a snag like an impactful risk or difficult change requests.

Competent Learners

The competent stage is when learners begin to see actions at least partially in terms of long-term goals. They are able to consciously and deliberately plan and can standardize processes. Competent learners start to see their actions as part of the bigger picture. They follow standards and guidelines but have an awareness of their specific situation. Competent PMs use planning to rationally anticipate scope creep, risks, and issues. They can use their previous experiences and develop plans to begin to identify risks and issues. Competent PMs can take on larger and more complex projects. They need additional guidance when the project hits a snag or when stakeholders need additional management.

Proficient Learners

Proficient learners see situations holistically rather than in terms of specific characteristics. They can assess a situation and identify what's important. They understand actions and their consequences. Decision making is less arduous, and they can more easily perceive deviations from the norm. Guidelines are used for guidance rather than direction. They are better able to make decisions based on information gathered, not just specifications and guidelines. Competent PMs can handle most projects with few mistakes. They may need additional support in the face of big changes or difficult situations. Proficient PMs involve stakeholders in most planning and decision sessions, while providing experience to help minimize risks. In addition, they know how to plan for risks along with how to generate risk mitigation plans based on both their previous experience and a fully holistic view.

Expert Learners

Experts are defined by Dreyfus as having intuitive decision-making abilities. They no longer rely on rules but instead have an intuitive grasp of situations based on a deep tacit understanding. They apply an analytical approach only when faced with a novel situation or when problems occur. Expert PMs use consistent processes and tools to manage complicated projects and teams. They intuitively know how a project should proceed and take a proactive stance to mitigate risks and avoid issues and scope creep. Expert PMs are the able to keep large, complex projects on track efficiently. See Table 3.1 for examples of characteristics and project management skills for novice, advanced beginners, competent, proficient, and expert learners.

Table 3.1 Novice to Expert Characteristics

Stage	Characteristics	Knowledge	Project Management Skills
Novice	• Rigid adherence to taught rules or plans • Little situational perception • No discretionary judgment	• Minimal, or 'textbook' knowledge without connecting it to practice	• Limited PM knowledge • Unable to plan for emergencies
Advanced Beginner	• Guidelines for action based on attributes or aspects (aspects are global characteristics of situations recognizable only after some prior experience) • Situational perception still limited • All attributes and aspects are treated separately and given equal importance	• Working knowledge of key aspects of practice	• Begin to apply previous experience • Cannot manage risks and changes well • Can follow standards and guidelines
Competent	• Coping with crowdedness • Now sees actions at least partially in terms of longer-term goals • Conscious, deliberate planning • Standardized and routinized procedures	• Good working and background knowledge of area of practice	• Able to use plans throughout the project to reduce risk and scope creep

(Continued)

Table 3.1   Novice to Expert Characteristics (Continued)

Stage	Characteristics	Knowledge	Project Management Skills
Proficient	• Sees situations holistically rather than in terms of aspects • Sees what is most important in a situation • Perceives deviations from the normal pattern • Decision making less labored • Uses maxims for guidance, whose meanings vary according to the situation	• Depth of understanding of discipline and area of practice	• Includes stakeholders in decision making • Uses plans to guide activities but able to flex as needed
Expert	• No longer relies on rules, guidelines, or maxims • Intuitive grasp of situations based on deep tacit understanding • Analytic approaches used only in novel situations or when problems occur • Vision of what is possible	• Authoritative knowledge of discipline and deep tacit understanding across area of practice	• Uses consistent practices • Functions proactively to prevent risks and scope creep • Manages project work across organization, just not within project

Learning is a continuum, and PMs can be anywhere on the novice to expert continuum. Novice PMs need more structure and guidelines and do best with smaller, less complex projects with well-defined standards and guidelines. As PMs gain more experience managing projects, they can be assigned to larger and more complex projects and can be expected to manage scope, budget, and resources more efficiently.

Chapter 4

Initiation – Novice

This chapter will cover the Initiation process group related to novice project managers and organizations without formal or mature processes. For many, the activities that are reviewed in this chapter are not well defined, communicated, or may not occur at all. Initiation begins with the request for a new project and ends when a decision is made related to the authorization of the project and when it will begin.

Project Requests

As mentioned above, the work that occurs during Initiation begins with the request. There are many types of requests that could include an identified need, an issue requiring resolution, a change in a regulatory requirement, or something more defined as with the desire to buy and implement new software. The requests could be submitted by any number of formal or informal ways. Below are a few examples of how requests can be received:

- Verbal – verbally requested when speaking with someone, in a meeting, while passing in the hallway, by stopping by the office, or in other formal or informal situations
- Email – an email sent to someone with authority, or access to someone with authority, to receive the request
- Request Form – completing and submitting a formal request form based on the defined process

- ■ Organizational Decision – senior leadership makes a strategic decision that includes one or more projects to meet, or support, the objectives and goals
- ■ Regulatory Mandate – a new, or updated, regulatory requirement has been identified, which requires some work to meet the expectations

Without a formal process, knowing who to send the request to may be confusing. Who has the authority to receive and act on a request? This may depend on the type of request and the formal and informal processes in place. The request could be related to information technology (IT), such as a implementing a new application or the need to expand the wireless network to cover areas where it is not available. The request could be department specific, such as developing a new process or a new training program. Or, the request could be a combination such as with a new patient care unit requiring new processes, training program, and the installation of hardware and network ports. With each of these examples, there should be someone identified to process the request or delegate this work. This may end up being the manager of the impacted department if not otherwise known.

Request Analysis

Once the request is received, it should be assigned to someone for analysis. This may be a member of the project management office (PMO) or someone from the impacted department. At the time the request is received, there is often very little known about the scope, size, or complexity of the work required. The effort may even be so small that it does not meet the organization's threshold to be considered a project. Depending on how much information is provided when the request was made, the requestor may need to be contacted for additional details. Information to be gathered from the requestor may include the following:

- ■ What is being requested?
- ■ Why is it important, or what is the justification?
- ■ What is the priority, or is there a deadline for when needed?
- ■ What is the budgetary impact and are the funds available?

Once the above information is obtained, the request can be further evaluated to determine what options are available for meeting this need.

Additional information can be gathered from other subject matter experts (SMEs) who may be from anywhere in the organization, depending on the specific request. The request information and the results of the analysis are documented for use later in Initiation. The type of information and format of this document would also depend on the request. There are traditionally three types of outputs from the analysis: a Change Request, a Project Charter, or a Business Case.

A Change Request is used when the request requires a small work effort and does not meet the threshold for a project. Most organizations have a change management process for making small updates to their applications. They may not have a defined process for non-IT changes that are smaller in scope and work effort. They may just assign them to someone to complete without a formal process.

A Project Charter is used when the specific work to meet the request is known. This is used to provide details for the work to be completed, the required resources, budget, milestones, and timeline. The solution must be known to be able to provide these estimates, although this document is typically high level and would be further refined during project planning, if approved. The Project Charter is further defined in Chapter 8 (Initiation – Expert).

A Business Case defines a business need and the available options to meet that need. This is used when the solution is not known, but the options have been evaluated to some extent. When this request is reviewed, the authorization decision could be made to move forward with detailed evaluation of the options with the expectation that the request would return with the outcomes and a justification for the recommended option. The Business Case typically includes the following information:

- Request/Project Name
- Version Number and Date
- Executive Summary
- Project Information
  - Submission date
  - Submitter and contact information
  - Business owner
  - Desired start date
- Project Description
  - Business need and benefits
  - Goals/Scope

   – Risks/Issues
   – Constraints/Assumptions
■ High-Level Business Impact
   – Business Functions/Processes impacted
   – Plans for ongoing operations
   – Justification and how it fits with organizational goals and objectives
■ Options and Analysis
   – For each option provide
      • Cost/Benefit analysis
      • Initial and ongoing costs
      • Return on investment
      • Security and privacy considerations
      • Resources required
      • Estimated timeline
   – Preferred option and justification for decision
■ Approvals
■ Appendix for references or additional documentation to support contents

Stakeholder Identification

During the analysis and development of the associated documentation, there will be stakeholders identified. They may be involved in the analysis or just provide answers to questions about the need. A project stakeholder is anyone who may be affected by the outcome of the project. This may be an individual such as the requestor, a group such as the staff who will work on the new patient care unit, or an organization such as the hospital itself. It is important to understand that the stakeholders include the project manager and project team, as well as anyone who believes they will be impacted by the project, whether true or not. Stakeholder identification begins during Initiation and is expanded on during Planning, and actually continues throughout the project.

One key stakeholder that should be identified during Initiation is the project sponsor(s). The sponsor(s) have overall accountability for the project and act as a link between the project team and the business. They have strategic-level decision-making authority and approve the project scope, any associated change requests, and the completion document.

They are also a reference for the project manager if anything requires escalation for resolution. There are times when the project would have more than one sponsor based on the scope or involved departments. For example, if the project is to implement a new software application for the pharmacy department, the project may have a technical sponsor from the IT department as well as a business sponsor from the pharmacy department.

Project Authorization

Once the information is gathered and documented, an authorization or approval decision should be made. The formal process for this includes a governance committee that evaluates requests across the organization and makes a decision based on the available resources (human, financial, and other) and the organization's strategic goals, objectives, and priorities. They consider all requests and realize the ones that are mandated either due to regulatory requirements, security needs, or necessary maintenance efforts. The governance process will be discussed further in Chapter 8 (Initiation – Expert).

With the absence of a formal governance process, the person or group who has the authority to approve the request may not be consistent throughout the organization. It may vary depending on the type of request, the different departments impacted, or if there is an IT component. In these situations, the request may be authorized without taking into account available resources or other requests across the organization. This could cause constraints related to the budget and available resources.

The final decision on the request could be any one of the following:

■ Approve Further Analysis of Options – this may include submitting a Request for Information (RFI) or Request for Proposal (RFP) or just further market research. This decision typically requires the results to come back to the approver with a recommendation and more details surrounding the project
■ Approve to Begin Now – this decision is made if the resources are now available or the priority is high enough that other work may be put on hold to free up the necessary resources

■ Approve to Begin at a Later Date – the decision is to move forward, but there is a need to delay the start. This could be due to unavailability of resources (human, financial, or other) or because of some predecessor that is required prior to beginning

■ Deny – if the decision is made to not move forward with this project, the justification for the decision should be documented

Once a decision is made, the requestor should be notified and all documentation should be stored in a location that is available by others. If denied, the documentation is archived for use as historical information if a similar request is received or if the request is revisited. If approved, the documentation should be made available for the project manager, when assigned. The Initiation process group ends once the decision is made, communication to the requestor happens, and the documentation is completed and posted in the proper location. Since the project manager is often assigned to the project after it is authorized, it may appear as if the Initiation process group did not occur, but is rare that these activities are skipped even if completed in a very informal way.

The typical reason the Initiation process group is not done occurs when communication about the project happens after it is already approved. These are projects that have a very high priority for the organization, are part of a mandate, or part of a larger organizational initiative. In these situations, the organization's leadership has already approved, or authorized, the project and communicates it should start right away moving it directly into Planning.

The Initiation process group begins with the initial request and ends with the approval decision, or authorization to move forward. Chapter 5 (Planning – Novice) describes the Planning process group. When the project is ready to begin and the project manager is assigned, planning begins. This is when the information gathered during Initiation is used to further define the project boundaries and how it will be completed. Chapter 5 (Planning – Novice) will discuss project planning for the novice project manager.

Chapter 5

Planning – Novice

Planning begins after the Initiation phase is completed and the project is approved. A quote often used when discussing project management is "if you fail to plan, you plan to fail." While there is no official author for this statement, some sites, including Goodreads (https://www.goodreads.com/quotes/), credit Benjamin Franklin with this quote. While this adage is a cautionary admonishment, the focus should not be on past failures, but instead on what lessons were learned with these failures. It is just as important to know what not to do as it is to know what to do. Planning is one of the most important roles of a project manager (PM). Successful projects have clear definitions of success at the beginning of the project with a clear end point. Many of us have been involved in projects without clear plans. Some of these projects may still have been deemed successful. Others may not have even been completed. Still others may have moved to completion, but without all of the planned outcomes or were delayed or over budget. Failing to plan adequately may not seem like a big deal in some instances, but in many healthcare projects, it can have a disastrous impact on the budget and outcomes. The bigger and more complicated the project, the more necessary planning is.

A well-planned project is more likely to be successful in terms of scope, budget, and timeline. After the Initiation phase has resulted in an approved project, the detailed planning can begin. The outputs and deliverables from the Initiation phase inform the Planning phase. During the Planning phase, project objectives are further defined and the detailed project management plan is developed. Planning fits into three general buckets – planning the success of the project, planning the actual project details, and getting buy-in from stakeholders. To increase the success of the project, questions that should be answered include:

1. Plan the success
   a. How will the project help the organization's strategic goals or priorities?
   b. What will happen if the organization does not do this project?
   c. What does a successful project look like to the stakeholders?
2. Plan the project
   a. What activities need to be done?
   b. What is the estimated duration of each activity and stage?
   c. What are the needed resources?
3. Get buy-in from stakeholders
   a. Are the stakeholders committing team members?
   b. Are the stakeholders committing to the duration and outcomes?
   c. Is the project approved by all required resources?

A major reason to plan is so that all the stakeholders have a shared understanding of the project and its expected outcomes, as well as to guide the execution of the project itself. Planning takes a fair amount of effort and should be included in the overall work and timeline of the project. While stakeholders, and even the sponsor(s), want the project outcomes now, remembering the adage about planning to fail helps focus on completing the planning before starting the work of the project.

Scope Definition

The first two steps in project planning – scope definition and gathering the team – can be done simultaneously or sequentially. Often having the team's input can help to refine a scope that is do-able and will reach the best outcomes. Gathering the team early, however, isn't always feasible if there is no work to be done, so at times, scope definition and other planning activities need to happen before the team is gathered. The PMI defines scope as "the sum of the products, services and results to be provided as part of the project" (PMI, 2017, p. 722). Another way to describe scope is that it is the features and functions that will be an output of the project. Still another way to define scope is to list the work to be performed to achieve specific outcomes. Scope defines what must be done and sets the boundaries for the project. Some scope definitions examples from our case study would include:

- Train staff
  - All staff working in the new Tower will complete computer-based training prior to working their first shift.
- Locate hardware
  - Gather requirements for hardware including network vs. Wi-Fi–enabled connectivity, size and weight of the devices, and number of devices available for the project.
- Review policies and procedures (P&Ps)
  - Review nine admission/discharge/transfer (ADT) policies (P&Ps) for potential changes due to the new Tower units.
- Revise the electronic medical record (EMR) for the new Tower
  - Update all unit, location, and service-type tables in the EMR.

Clearly defined scope is necessary to successfully complete any project. Clearly defined scope also helps the stakeholders know that the project was completed and successful. Scope, in the end, is what is expected by the stakeholders by the end of the project. Scope definitions should include what *is* and what *is not* part of the project. In our hardware example, for example, negotiating prices for the equipment is not part of the scope of the project; rather it will be done by others outside this project. Well-defined and agreed-upon scope helps prevent 'scope creep', the bane of any project. Scope creep is defined as adding unauthorized features or functions (Larson & Larson, 2009). Scope creep can be unavoidable and happens most frequently because of unclear or changing requirements, or because a stakeholder is looking to do more than what was already agreed upon. Often, scope creep happens because of good ideas but still impacts the project and needs to be managed. One of the best ways to manage scope creep is with a change management plan that is addressed in Chapter 9 (Planning – Expert). In projects of short duration or those that are less complicated, scope creep can often be managed by considering the requests a future project.

Managing scope is important for a number of reasons. First, the project is intended to accomplish specific outcomes or goals within a given period of time, and often with a specific amount of resources. Some projects also plan to see costs reduced as a result of the project, and budgets are adjusted accordingly. Any increase in scope can adversely impact costs, time or resources, and, more importantly, the quality of the work. Called the triple constraint, the relationship between scope, time, and resources, is often what the PM spends most of their time on during the project. (More about the triple constraint in is covered in Chapter 9 [Planning – Expert].) By defining scope and gaining stakeholder buy-in, scope creep can be minimized. Scope should be defined in ways stakeholders understand. Some examples from our case study include:

- Train all staff prior to the Tower opening
- Identify locations for 20 new PCs on each unit
- Review nine P&Ps related to admission on the unit
- Develop test scripts for admission, discharge and transfer of patients

Scope will drive the workplan development and resource requirements.

Gathering the Team

Having the project team available to help finalize scope can improve the likelihood that the scope is complete and do-able within the time allotted. Human resources are often the costliest and most difficult resources to manage part of any project. Even if the 'team' is one person, other work that person is responsible for needs to be adjusted so that they can focus on the new project. In addition, larger teams may come from different departments and look at the project from differing points of view.

The first step in managing the team is to clearly define roles and responsibilities of each team member – including what they will and will not do as part of the project. Role descriptions for the PM or lead, sponsor(s), stakeholders, advisory groups, as well as the analysts, testers, record keepers, and content experts are needed. Clearly defining everyone's area of responsibility will help to reduce confusion and overlap among the team. This can be a simple table with the role and basic responsibility tasks such as in the Table 5.1. It can also be a resource management plan that we will discuss further in Chapter 9 (Planning – Expert).

Many of these resources can help formalize the scope and should agree (in writing if possible) to both the final scope and their role in the project. Once the final scope is approved and the team's roles and responsibilities are defined, the project timeline can be created.

Table 5.1   Sample Project Roles and Responsibilities

Role	Responsibility
Sponsor(s)	• Champion the project • Secure project funding
Stakeholders	• Commit user resources • Communicate business needs
Project manager	• Create and drive project timelines • Identify and manages project issues and risks • Direct work efforts of project team toward milestones
Team members	• Complete all assigned project-related tasks and activities • Identify and reports issues • Update project manager on work progress
Trainers	• Train staff • Assist at go-live

Project Timeline

The project timeline drives the work that needs to be completed in order to attain the scope agreed upon by the stakeholders. Project timelines can be as simple as a few steps. For example, if your project is developing training materials, the steps of your project could include:

■ Project approval
■ Gather team (resources)
■ Identify new processes and P&Ps (scope)
■ Create timeline (schedule)
■ Develop training materials
■ Deliver training
■ Support go-live
■ Close project

Another project example is to review P&Ps for the new units. Tasks for this project could include:

■ Determine which P&Ps need to be reviewed (scope)
■ Determine who will review each P&P (resources)
■ Communicate due dates to reviewers (schedule)

■ Review P&Ps
■ Approve revised P&Ps
■ Work with the trainers to develop training (if needed)
■ Publish new P&Ps
■ Close project

In this example, finalizing the detailed scope and gathering the team is part of the project timeline since the actual project has already been approved. In addition, revising P&Ps is usually done on a schedule, but this may require an additional review because of the new Tower.

There are many names for the project timeline. Some PMs call it a Work Breakdown Structure (WBS). The PMI defines the WBS as a "hierarchical decomposition of the total scope of work to be carried out by the project team to accomplish the project objectives and create the required deliverables" (2017, p. 161). A WBS is a visual representation of the tasks. It often looks like an organization chart as the tasks and categories are developed. See Figure 5.1 for a basic WBS example.

Once all the tasks are created and due dates added to the basic WBS, PMs may call it a timeline, and when created in project management software, they may call it a workplan. Another way PMs refer to the timeline is called the schedule. Usually a schedule is a calendar view of the tasks, rather than a linear view in the workplan or timeline. Whether called a workplan, timeline, schedule, or WBS, it should include, at a minimum,

■ A list of tasks (scope)
■ Start and end dates for each task (duration)
■ Responsible person or persons for each task (resource)

Creating the workplan, as it will be called going forward, is an art and a science. There are many ways to get started creating a workplan. One way

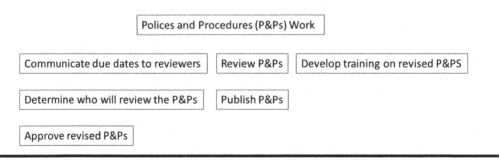

Figure 5.1 Basic WBS.

is to document every known task with expected duration and responsible person or group. This method, however, assumes all the tasks, how long they will take, and who the best person is to successfully accomplish the task is known. Unless you are a very experienced PM or you are duplicating an existing project, however, this method is extremely difficult. When starting out as a PM of a project, there are many easier ways to plan the workplan.

As a novice PM creating a workplan, it may be easiest to just list tasks, without duration or resource assignments, as they come to mind. Brainstorming a list of tasks without taking the time to organize them in order initially allows the PM to focus on what needs to be done rather than getting hung up on how long it would take or who the best resource would be. One way to brainstorm a workplan is by writing (or typing) each task identified on a sticky note – one task per note. This can be done on paper or electronic sticky notes or other building block software. The benefit of using one sticky note per task is that once all the tasks are listed, they can then be organized into groups or in the order they need to be done. Another way to capture a list of tasks is to type each task in a word processing table or spreadsheet. After all the tasks have been captured in the table, they can be copied and pasted into groups or in a different sequence. Whiteboards, whether static or smartboards, could also be used to list tasks. Once all the tasks are listed on the board, you can designate grouping with lines or colored markers. For example, all the preparation tasks could be circled with one color and the training-related tasks in another color.

No matter which method used, few PMs can be expected to identify every single task. It is often helpful to pull others into the process. Team members who will be doing the work may have suggestions about tasks needed to complete the work. Other PMs may have suggestions based on projects they have managed. Selected stakeholders may be helpful as they know the business that will be impacted by the project. The PM could interview team members or stakeholders to gather more information about tasks needed. Team members and/or stakeholders could also be gathered together for a workplan planning session to further detail the work.

Once all the tasks are listed, the next step is to organize them. Grouping tasks by resource or phase helps to organize the work and can even help to identify additional or even unnecessary tasks. Key phases of any project should include:

■ Analysis (plan)
■ Build

- Test
- Train
- Go Live
- Post go-live support
- Closure

More detailed projects may have additional phases, but almost every project has tasks in each of the phases listed above.

The next step for the workplan is to add durations and resources to each task. Often, duration is an estimate that gets refined as all the steps are resourced. For example, if the committee reviewing the P&Ps only meets once a month, can only review three P&Ps per meeting, and there are nine P&Ps to be reviewed, then the total duration for the review task is 3 months. This may be too long for the project to be completed in time and may need to be reduced. Remembering the triple constant, there are three options to reduce the duration of this task. The first option is to reduce the number of P&Ps to be reviewed (scope). The second is to have the committee meet more often (resources) or extend each meeting time (schedule). Tasks should be included in the workplan based on the approved scope, then adjusted as needed to reduce the overall duration or number of resources. In Chapter 9 (Planning – Expert), we will discuss how to determine how many tasks should be included in your workplan.

Assumptions and Constraints

Assumptions and constraints should be created in conjunction with the workplan. The PMI defines assumptions as "a factor in the planning process that is considered to be true, real or certain, without proof or demonstration" (2017, p. 699). They define constraints as a "limiting factor that affects the execution of a project" (PMI, 2017, p. 701). An assumption in our project could be that all nine ADT P&Ps will be reviewed and updated as needed. A constraint could be that only the admission committee can review these P&Ps. Assumptions and constraints can be identified throughout the Initiation and Planning phases. Once defined, they are used as supporting documents to the workplan to help control the project work and scope. Any changes to the assumptions or constraints would be managed by the change management plan (more about this in Chapter 9 (Planning – Expert)). Examples of assumptions for our case study could be:

- Staffing ratios will remain the same in the new unit
- 22-inch monitors will be mounted on each mobile cart
- Nine ADT P&Ps will be reviewed
- New equipment will be tested by IT prior to deployment

Examples of constraints for our project could include:

- The new Tower will open as scheduled
- Staff from other units are not able to float to the new unit
- Hardware must be positioned on existing cart configurations
- No additional staff will be added during the move
- Hardware will not be deployed until the week before the opening
- All patients must be moved into their new beds by 11:59 PM the day of the move

Assumptions and constraints should be elicited from stakeholders and the project team. Without clearly defined assumptions, work could expand and derail the whole project. When a third party is involved in a project, assumptions and constraints are often included as part of their contract.

Budget Planning

In small, less complex projects, the PM is usually not responsible for managing a budget. Often, any funding for the project is absorbed in the day-to-day operations. The PM, however, should track project expenditures when they occur. At times, this is just tracking resource time. At other times, this is a basic spreadsheet with a list of all resources used (human and technical), the cost, and a total. For example, for the P&P case study, the only cost will be the resources who will review the P&Ps as part of their normal job requirements. Tracking the amount of time these resources spend on the project is probably enough to manage the budget. For the training case study, there may be more data that needs to be tracked, although this may not be the PM's responsibility. For this example, the PM needs to track the trainers' time to develop the new materials, the cost of any software needed, and staff time to attend training. Salary estimates for each type of trainer and learner are usually used to track these costs in a small project. See Chapter 9 (Planning – Expert) for more detailed budget management.

Communication Plan

Once the scope statement has been developed, the team has been gathered (if any), and the workplan with assumptions and constraints has been created, it's time to develop the communication plan. The purpose of the communication plan is to define the five rights of communication which are:

1. The right information
2. To the right people
3. At the right time
4. In the right format
5. Stored the right way so that it can be used for reference during and after the project

During the planning stage, stakeholders are often excited about the prospect of a successful project. This can make discussing their potential communication needs easier. It is much better to have conversations about how to manage potential issues before they occur than in the heat of the moment. Sharing positive progress can be communicated in an email or posted on the project page. Issues or delays, however, may require more information or a face-to-face meeting. While it is often said that you "can never communicate too much," planning to communicate the appropriate information with the appropriate frequency is key. PMs need to first determine who needs information about the project. At a minimum, stakeholders, team members, and the project sponsor(s) need to know when the project kicks off, the overall workplan, and anticipated outcomes. Some of this information is part of the project charter that was drafted during the Initiation phase. In the Planning phase, this information is further refined through the detailed scope statement and workplan. Stakeholders and team members need to be informed about these details at the start of the work and throughout the project.

The second part of the communication plan is including the right people. Determining who needs information about the project is necessary. Answering questions such as those listed below will determine who needs to be included in the communication plan.

- Who will be impacted during the project?
- Who will be impacted by the project outcomes?

- Who will be impacted if the project is a success?
- Who will be impacted if the project is not a success?
- Whose work will change as a result of the project?
- Who will do the project work?
- Who is funding the project (including the resources)?

Short, simple projects have fewer stakeholders who require project updates. Some quality assurance or evidence-based projects may only require updates at the end of the project. Longer and more complex projects need more updates. Reporting status at the end of each phase – planning, build, test, train, go live, post go live, and closure – keeps all stakeholders informed throughout the project. Both simple and complex projects can follow a similar communication plan, but longer projects need more updates. A month-long project may need only two to four updates. A 3-month project may have the same number of updates; however, they will be spaced differently. For example, project updates can be delivered:

- Project kickoff
- Prior to starting build
- Prior to starting testing
- Prior to starting training
- Just before go live
- After the first outcomes are measured
- At project closure

Longer projects usually follow a weekly or monthly schedule, with added updates at key periods in the project (if needed).

Status updates can be delivered in a variety of ways. Some methods include:

- Face-to-face meetings
- Written status reports
- Email updates
- Posts on the project site
- Newsletter articles
- Electronic dashboards
- Verbal or written reports at regularly scheduled department or stakeholder meetings

The delivery method should match the stakeholder needs and the urgency of the information. Regular updates can be written and distributed either directly to stakeholders via meetings or emails or posted on a project sites. Updates with minor issues or when simple decisions are needed are better delivered in a meeting or as a push notification with clearly defined decision parameters. For example, an email can be sent to stakeholders using a format similar to SBAR (Situation-Background-Assessment-Recommendation). After describing the decision needed (situation) along with supporting information (background), PMs can describe the options with consequences (assessment). Items requiring a recommendation should then be listed so stakeholders are clear about what they are recommending. In many email packages, a voting button can be added to the email to capture each stakeholder's recommendation. In addition, there are many free applications to collect responses. When recommendations are requested of stakeholders, the date the recommendation is needed should also be included in the email or presentation. In addition, many PMs include verbiage to the affect that no response means the first option or agreement on the recommended option. Gathering the resulting votes and reporting the outcome to the sponsor(s) for final decision will close the issue, and it can then be added to the workplan or marked resolved in the log.

Major issues, however, often require a more interactive discussion. The SBAR method will still work, but for major issues, stakeholders may want to understand how other stakeholders view the options and/or offer alternatives. Discussing issues at a status meeting and adding them to the status report is usually enough, but additional meetings may be needed. As with all status updates, the information should be documented and stored with the other project documents. The PM is responsible for maintaining all communication about the project. In short simple projects, information is often stored on the PM's computer or cloud drive (following the organization's storage polices). Larger or more complex projects may have shared storage such as SharePoint or a project organization page. All communication should be stored until the project closure phase is complete, and then it can be archived. More about project closure will be discussed in Chapter 7 (Closing – Novice) and Chapter 11 (Closing – Expert).

Often it is useful to create a grid with the five rights of the communication plan. The communication model is defined as "a description, analogy, or schematic used to represent how the communication process will be performed for the project" (PMI, 2017, p. 700). Seeing the information graphically can help all interested parties understand what communication about the projects they can and should expect. This communication grid can also help to determine holes in the plan or resources that are getting too much or too little information. See Table 5.2 for an example of a communication grid.

Table 5.2 Communication Grid

Stakeholder	Information	Frequency	Format	Storage
Sponsor	Project Status, Outcomes, Issues, Risks, Budget	Weekly/ Monthly*	Face-to-face Written** Posted**	Written reports stored on project site
Head of Impacted Department	Project Status Outcomes	Weekly/ Monthly*	In person and in writing	Written reports stored on project site
Other Department Heads	Project Status	Beginning of Phase	Newsletter Staff Meeting	Newsletter articles and reports to staff meetings stored on project site
Providers	Project Status	Beginning and End of Project	Department Meeting	Reports to department meetings stored on project site
Clinicians	Project Status	Beginning and End of Project	Newsletter Staff Meeting	Newsletter articles and reports to staff meetings stored on project site
Team	Project Status Issues Risks	Daily/Weekly*	Face-to-face Written** Posted**	Written reports stored on project site

* Frequency depends on project duration and complexity.
** Written and posted format supports what is presented in face-to-face meetings.

Once the scope, resource plan, workplan with assumptions and constraints, and communication plan are completed, the PM should get written sign-off from the project sponsor(s). This formality ends the Planning phase and increases the likelihood that the project will follow the agreed-upon plans. More information about detailed status reporting, success factors, and sign-off will be addressed in Chapter 9 (Planning – Expert). Once the Planning phase is complete, it is time to kick off the project.

Chapter 6

Execution, Monitoring and Control – Novice

Now that project planning had been completed, the full project plan has been developed, and the team gathered, it's time to start the execution and monitor and control phases of the project. The execution phase is where all of the activities defined in the project management plan are completed. Its where the work of the project gets done. It includes 'build', test, training, and go-live activities. The Monitoring and Control phase is simultaneous with the Execution phase and includes the main work of the project manager (PM) during these phases. The purpose of the PM's work during the monitor and control phase is to make sure the work gets done on time, on budget and to scope. The execution and monitor and control phases will be referred to as one phase throughout the book. During this phase, the PM will focus on managing the work, controlling issues, risks, and change request, and assuring that project quality is met.

One way to think about the PMs' role during the execution phase of the project, is that the PM functions like an orchestra conductor. The PM's job is to guide the team to complete the scheduled work on time, within budget and to meet the project requirements and scope. A conductor, like a PM, needs to know the musicians' strengths and abilities (team), the music to be performed (scope), when the performance is scheduled (workplan), and the best way to present the music to the audience (stakeholders) in the venue where they will be performing. While a conductor doesn't have to know how to play every instrument in the orchestra, they must know how to read music and the arrangement to be performed. During their first practice (kickoff), the conductor will explain the plan to the musicians, clarify their roles (i.e., first or second chair, etc.), and expect them to begin to practice (the work) in order to correctly perform the music. Then the conductor monitors their progress toward the performance through practice sessions (status meetings). While the musicians are expected to correctly perform the music, the conductor's job is to monitor their work and identify any issues or risks that may adversely impact the performance and assist them to resolve the issues. A conductor also needs to understand the venue set-up and acoustics (culture) in order to maximize the performance. In this way, the conductor can increase the likelihood of a flawless performance.

The PM's work during this phase is similar. While the PM is not expected to know how to do all the work of a project, they do need a basic understanding of the work needed, how much time it should take and how to assess for potential issues and risks. By monitoring and controlling the project, the PM increases the likelihood that it will be completed on time, within budget and to the stakeholder's satisfaction. PMs will use all the tools developed during the Planning phase to monitor the work against the plan, report to stakeholders, manage issues and risks that crop up during the project, and to manage the team.

Project Kickoff

One of the first steps in the Execution phase is to make sure the team members and stakeholders understand the project and familiarize them with details of the plan. Most projects start with a kickoff meeting that includes key stakeholders as well as the team members. Some PMs consider the kickoff meeting as the end of the Planning phase. Others use it as a driver for completing detailed planning. Either way, kickoff meetings set the tone for a project and inform all participants of the key activities and timelines. Topics to include in the kickoff meeting include:

- Team introductions
- Project purpose and outcomes
- Project scope
- Key activities in the workplan
- Processes that will be used to communicate project updates, issues, risks, and change requests

The kickoff meeting should begin with an introduction of all of the key team members and stakeholders. Each person should tell a little about their specific role in the project. This way, each team member and stakeholder can have an idea of who is involved with the project. This may be the only time everyone gets together, so it is important to set the stage for the project. For example, everyone should know who the project sponsor(s) is. The sponsor(s) helps to drive the overall direction and value of the project.

The kickoff meeting is a good opportunity to make sure stakeholders are aligned on the purpose and scope of the project. For smaller, less complex projects, a kickoff could be part of a regularly scheduled meeting. For example, if part of the project is to collect data in a specific unit, the kickoff meeting could explain why the data is needed and what the expected outcomes are, and tell the staff that someone will be collecting data or reviewing charts or asking them questions, etc. In our case study example of developing training materials, the kickoff may include information about the future unit so everyone can think about what training will be needed in order to take care of patients in the new unit configuration or equipment that may need to be replaced or increased in number. For the policy and procedure project, one of the kickoff activities could be to explain how team members are going to monitor existing admission, discharge, and transfer processes in order to potentially revise policies.

Presenting the scope, including what will and will not be included, in the project is one of the main reasons for the kickoff. It is important to describe the scope in terms that will resonate with stakeholders. Rather than presenting detailed tasks of the project, scope can be presented as how the work will attain the specific outcomes. In our case study example, scope presented to the stakeholders could be:

■ Nine ADT policies and procedures will be reviewed and updated in advance of moving patients into the new unit
■ Staff will be oriented to the new unit using materials developed during the project from the existing training packets
■ Training materials will be available online for every staff member to complete
■ Hardware will be installed in the new Tower prior to the go-live
■ Time to treatment will be analyzed 1 week before and 1 week after the new unit opening

The PM will need to address stakeholder questions in regards to scope during the kickoff meeting and often throughout the project. It is important to be clear and firm when discussing what scope was agreed upon. Including the assumptions and constraints that formulated the project management plan, team makeup, and workplan when addressing scope can help solidify agreement.

Reporting Project Status

Another purpose of the kickoff is to model status reporting. Much of the success of any project is how the stakeholders believe the project is going. By modeling status reporting during the kickoff meeting, stakeholders will know what they can expect throughout the project. In a status report, the PM should share information about progress against scheduled activities (tasks), planned next tasks between now and the next status report, issues including plans to address them, and any anticipated risks during this phase of the project. Status reports should include at a minimum:

■ Symbols or words to describe overall project status against the project plan (e.g., a red-yellow-green stoplight)
■ Issues and risks for stakeholder information (or that they need to address)
■ Change requests
■ Plans for next period

One example of a status report may show a stop light icon about the project with definitions to define what red, yellow, and green lights mean in relation to the project. Often, green means that milestones are or target, ahead, or less than 5% behind. Yellow often means tasks are 5–15% behind or performance indicators are below baseline, but the project is expected to catch up and at this point not delay the timeline or increase the budget. Red usually means that the project is in trouble with missed milestones or a critical path that is more than 20% behind. When a project is in the 'red', the project is often at risk for missing the go-live, budget, or scope requirements if something is not done. Identifying the status indicators of the project during the kickoff will make further communication with the stakeholders clearer and more effective. The legend should be included as part of the status report so that stakeholders can see the meaning of the colors in

Color	Indicator	Definition
Green ✿		Milestones are on target or less than 5% behind Performance indicators are on target
Yellow ✿		Milestones are 5 – 15% behind Performance indicators are below baseline
Red ✿		Critical milestones have been missed or are greater than 20% behind Performance indicators are below baseline and may not be able to be fixed

Figure 6.1 Status report stoplight descriptors.

the report. Any indicators should include both color and some other method (such as fill) so that people with disabilities, such as color blindness, can see the indicators. See Figure 6.1 for an example of a stoplight table legend.

Status reports should include information about the tasks that were scheduled to be completed by this time in the project and tasks scheduled in the next phase. During the kickoff meeting, few, if any, tasks should have been completed, so the majority of the report will be focused on planned next steps. General information about the tasks such as the focus of the tasks, anticipated risks to completing the tasks, and any mitigation plans to reduce the likelihood of the risks becoming issues, should be addressed. If there are any issues that need to be addressed now, they should also be described. Details about each task, who will complete it, how it will get completed, etc., are not needed. The stakeholders only need enough information to help them determine what, if anything, is needed from them or their department during the next phase and how they can help mitigate any risks.

During the kickoff meeting, some stakeholders may be hearing about the project risks for the first time. Risks are a normal part of any project, but not all stakeholders will know this. Only risks identified during the planning phase and those that may happen during the early stages of the project should be shared during the kickoff meeting. In addition, it is important to always include the risk's mitigation plan when discussing risks. For example, in our project to review and potentially revise nine ADT policies, a risk could be that there are additional policies that may need to be revised as a result of reviews of the initial nine. One way to mitigate this risk and prevent it from becoming an issue that could impact the project could be to make an assumption that any additional policies identified would be reviewed and revised after the go-live. Doing this early in the project would allow time to plan any policy changes directly after the initial go-live. The purpose of discussing risks during the kickoff meeting is two-fold. First, stakeholders need to know that risks exist as they do in any project, and that a mitigation plan has been developed for each identified risk. Second, stakeholders may identify additional risks and can help identify mitigation plans.

Part of the kickoff meeting should also describe the change control process. Requests to add to or change planned scope frequently occur. The change request process should include a method to track requests, analyze the potential benefits and impacts, and who will make the final determination of the requested change. Describing this process early in the project will help make the project change process smooth. The kickoff meeting should end with a discussion of when and how stakeholders will get communication about the project. In a small or less complex projects, this may be another announcement at the next regularly scheduled meeting or an email update. For a larger or more complex projects, regular status meetings and schedules for written reports should be established and follow a predefined schedule. They could be scheduled as often as weekly or as infrequently as monthly, as long as they follow a predefined schedule so that stakeholders know when to expect information. Status meetings could be scheduled monthly on the third Monday, for example. The PM should schedule these in advance of the kickoff meeting and share the schedule, location, dial-in information (if applicable), and expected attendees as a way to close the status meeting. In addition to holding status meetings, the PM should follow the other steps in the communication plan throughout this phase. More information about the communication plan is available in Chapter 5 (Planning – Novice).

Monitoring the Work

Throughout the Execution phase, the PM monitors the project work against requirements and the project management plan. For example, the PM needs to review the P&P committee's progress on the nine ADT P&Ps. The PM needs to understand if the committee reviewing the nine ADT P&Ps is doing that exclusively or if they have other work at the same time. In our example in Chapter 5 (Planning – Novice), we determined that the committee needs to complete three P&Ps per month. If the PM did not plan for simultaneous work in the committee, this may cause the project to be delayed or require additional resources to complete the review. Assessing and reporting progress weekly or even more often in a shorter project and when the project nears completion is key to keeping the project on target.

According to a well-known project management tenant called Parkinson's law (1957), the work typically fills the time allotted. If the team member is given an hour for a task, it will usually take that hour. If they are given a week to do a task, the task will not be completed until the end of the week. This is why the 'busiest man' tenant exists in conjunction with Parkinson's law; an individual will theoretically deliver in the shortest amount of time when they have the least time to spare. In order to increase the likelihood that the P&Ps will be done on time for the new unit opening, the PM needs to break down the task into manageable chunks for the committee in order to combat Parkinson's law. Then the PM needs to monitor the work closely, not just capture that the work is completed. Rather than saying that three P&Ps need to be completed each month, the PM needs to calculate working days and time available. Since there are only about 20 working days in a month, in order to complete three P&Ps in 1 month, each P&P needs to be completed in about 6 working days. Often, team members will work on more than one task or project at a time, which can make it harder to assess progress against the timeline, as the team member may decide to bundle like project tasks together and not work on just one assignment each week. This can make project delays more likely if the PM does not adequately monitor the work. The PM needs to work with the team to understand the progress of the work, and then report it and any issues to the stakeholders as they arise.

The team is focused more on getting the current tasks done, but the PM is often focused on the next step or phase of the project in order to keep the project moving forward. For example, as soon as the project kickoff is complete, the PM should begin to assess for issues and risks or other potential delays or roadblocks and make sure the next tasks can be successfully completed. Training tasks, for example, often require that other 'build' steps be completed before they can start. The PM needs to work with the training team to determine the minimal 'build' effort needed in order for them to start their tasks, as well as monitor their activities to be sure that they are doing tasks that are not linked to other 'build' activities. The training team could set up classrooms, learning management systems, curriculum maps, classroom policies, processes for remedial training, etc., for example, while waiting for the actual 'build' to progress. Since the PM is focused on the next activities, they can have a better view of the work that can progress and should direct the teams accordingly.

The PM should update the workplan regularly throughout this phase. At a minimum, updating the workplan should include marking tasks complete. Each task should have a due or end date, and once it is completed, the task status should be updated to a status of complete. Just marking time, i.e., crossing off days, is not managing the workplan. The focus should be on task completion, not time progression. While marking tasks complete is the minimum way to update the workplan, however, it does not give the PM much useful information. In addition to the end or due date, tasks should also have a start date. The PM should be monitoring tasks to see that they start as scheduled and updating the workplan accordingly. Tasks could be updated to "in progress" or "started" to clearly identify which tasks are on schedule even before they are completed. Color coding can also help to easily show the project status. Tasks that are on schedule, those that are completed, in progress as planned, or not scheduled to start yet, could be colored green. Tasks that are behind schedule but expected to catch up can be colored yellow. Tasks that should have started but have not and tasks that are behind and not expected to catch up can be colored red. If possible, the same criteria and legend as defined for the status report should be used to code workplan status. See Table 6.1 for a sample workplan in progress

In an ideal world, each task would be completed as scheduled. In reality, some task will not start as scheduled or will take longer than originally planned. If either of these issues occurs, the workplan should be updated. The PM should update the start or end date if they are not the original dates. In a simple or very short project, this can be done by changing the original dates. It is better and more useful, however, to add additional columns for the actual start or actual finish dates. In this way, it is clear to all exactly what is happening with the project and where the main issues were. See Table 6.2 for an example of a workplan with planned and actual start and end dates.

The PM will continue to update the workplan until all the tasks are completed and the project is completed. Chapter 10 (Execution, Monitoring and Control – Expert) will address more complex project workplans and the use of project management software.

Managing Budget

At the same time the PM manages the workplan, they should also track monies spent on the project. In smaller, less complex projects, the PM may

Table 6.1 Sample Workplan in Progress

#	Task	Start Date	End Date	Responsible	Status
1	Determine which P&Ps need to be reviewed (scope)	1/1/2020	1/1/2020	Clinical Team	Complete
2	Determine who will review each P&P (resources)	1/2/2020	1/6/2020	P&P Committee Chair	Complete
3	Communicate due dates to reviewers	1/7/2020	1/7/2020	PM	Complete
4	Review P&Ps	1/13/2020	3/13/2020	P&P Committee	In progress
5	Approve revised P&Ps	2/13/2020	3/13/2020	Steering Committee	In progress
6	Develop training	2/13/2020	3/20/2020	Trainers	Not started
7	Deliver training	3/23/2020	4/10/2020	Trainers	Not started
8	Publish new P&Ps (go-live)	4/10/2020	4/10/2020	P&P Committee Chair	Not started

not be responsible for much of the budget but still needs to track against the plan. Tracking resources' time in the workplan can be a part of tracking the budget. For example, if according to the workplan, the P&P committee should spend 8 hours per policy, there are 10 committee members and there are 9 polices, then the total P&P committee should spend at least 720 hours on this project. If they complete their tasks on time, the PM often assumes they spent the allotted time on the project. In larger, more complex projects, resources are more likely to be required to report actual time spent.

In addition to tracking resource time spent, the PM will need to track monies spent on hardware and software. As hardware is delivered, the PM should document against the plan. Costs for software or other tools needed to complete project activities should also be tracked against the planned budget. As long as the project is on budget, the status report can include one line stating this for the stakeholders. If the project is not on budget, this should be included as an issue and managed accordingly.

Table 6.2 Sample Workplan with Actual Start and End Dates

#	Task	Start Date	End Date	Actual Start Date	Actual End Date	Responsible	Status
1	Determine which P&Ps need to be reviewed (scope)	1/1/2020	1/1/2020	1/1/2020	1/1/2020	Clinical Team	Complete
2	Determine who will review each P&P (resources)	1/2/2020	1/6/2020	1/2/2020	1/8/2020	P&P Committee Chair	Complete
3	Communicate due dates to reviewers	1/7/2020	1/7/2020	1/9/2020	1/9/2020	PM	Complete
4	Review P&Ps	1/13/2020	3/13/2020	1/13/2020		P&P Committee	In progress
5	Approve revised P&Ps	2/13/2020	3/13/2020	2/13/2020		Steering Committee	In progress
6	Develop training	2/15/2020	3/20/2020			Trainers	Not started
7	Deliver training	3/23/2020	4/10/2020			Trainers	Not started
8	Publish new P&Ps (go-live)	4/10/2020	4/10/2020			P&P Committee Chair	Not started

Managing Scope

One of the most difficult roles the PM has during the Execution phase is managing scope. Scope, as defined in Chapter 5 (Planning – Novice), is the work that needs to be done in order to achieve the outcomes. Each team member should have a clear understanding of what is and what is not in the agreed-upon scope. The PM should discuss scope with the team so they are clear before the kickoff meeting. In this way, the scope becomes a living, breathing part of the project with the team in agreement on what will be accomplished. In small, less complex projects, managing scope can be simple but still needs to be done. What can seem like a small request for a change in scope can quickly escalate. For example, if the project plan defined that nine ADT P&Ps would take 3 months to complete, adding even one more could derail the whole project. It could potentially delay the opening of the new unit, require experts to be pulled off other work in order to review the additional P&Ps, or allow the go-live without clearly defined P&Ps. In each of these scenarios, there is a potential negative impact to patient care and the organization's bottom line. Managing scope reduces the possibilities of delays, overruns, or reduced quality.

In order to manage scope, the PM has to analyze every change request for the potential impact to the project timeline, resources, and expected outcomes. This analysis should include input from the team as well as the stakeholders. Members of the team should be able to estimate the impact of the change on the work – the timeline and the resources. Stakeholders should be able to estimate the impact of the outcomes if the requested change is not added, as well as the potential benefits if the change is added to the project. Once the PM has gathered all of the information together, they then present the options with pros and cons of each option for a recommendation to take to the sponsor(s) for approval.

Change Requests

PMs for a simple project still do many of the same activities as a PM in a more complex project. The activities themselves may be fewer or less complex, but the execution and control work remains mainly the same. PMs always need to manage scope, report to stakeholders, and manage issues and risks. In a less complex project, there is less chance for revisions in scope. Activities are carried out by fewer team members, and there are usually fewer stakeholders involved. Often requested additions in scope can be delayed until after the project is completed or during the next phase. In more complex projects, however, requests to change scope need to be analyzed against the impact to the overall project. As discussed in Chapter 5 (Planning – Novice), any change in scope could impact the timeline or resources (both people and budget). The PM needs to determine how much effort is available to analyze the potential changes in scope. If it is obvious to all that the change request is not do-able within the existing project scope and timeline, often the PM can just deny the change request, although that is rare. More often, the PM will need to analyze the requested change request against the project timeline and resources to determine what options are available. For example, if a key stakeholder requests the change in scope, the PM must determine the impact to the project of doing the added scope and the impact to the overall success of not doing the added scope. The PM may need to pull in other team members or business owners to assess the request. This takes time, and potentially, in and of itself, could impact the project timeline. Often PMs build time into the project plan to address scope change requests, but usually these requests are not planned.

Once the analysis of the impact of the change request is complete, the PM will need to present it to the sponsor(s) for a decision to include or not include the requested change. When presenting to the sponsor(s), the PM should include the pros and cons in terms that resonate with the stakeholders. Pros and cons should include impacts to the timeline, resources, and projected outcomes of the project. Once the sponsor(s) makes the decision about the change request, the change request should be documented, along with the decision, in the change control log and the PM should share the decision with the stakeholders. If the scope is added to the project, the PM will need to revise the scope document and the workplan accordingly and re-baseline the project. The most important thing about requests to change scope is that they are analyzed against the project and recorded in a log along with the final decision. Change requests often come up again if the requested feels strongly about the change. The request should only be re-analyzed if new information has come to light. There are times when a change request needs to be carried out in order to successfully complete the project and meet the expected outcomes. Sometimes work was not adequately planned or known. As with any change in scope, the workplan, issues, and risks logs should be assessed and updated as needed.

Managing Risks

Issues and risks are often managed together during a project but are very different. PMs often intermingle issues and risks, but they should be managed differently. Issues are defined as "a current condition or situation that many have an impact on the project objectives" (PMI, 2017, p. 709). In order words, an issue is something that will happen or should have happened but didn't and will impact the project's timeline, scope, or outcomes if not resolved. Risks, on the other hand, are defined as "an uncertain event or condition that, if it occurs, has a positive or negative effect on one or more project objectives" (PMI, 2017, p. 720). A risk is a possibility – it is something that may or may not happen but could impact the project timeline, scope, or resources. Some risks are identified during the project planning phase. Others may have been identified during the kickoff meeting. Still others could have been identified during the project execution. Risks need to be managed to prevent them from becoming issues. Whenever a risk is identified, a mitigation plan should be developed. Risks should be maintained in a log. At a minimum, the risks log needs to include:

- An assigned risk number
- Description of the risk
- The date identified
- Likelihood of the risk occurring (high, medium, low, for example)
- Expected project impact if the risk does occur (high, medium, low, for example)
- Mitigation plan

A mitigation plan includes ways to prevent the risks from happening. For example, using the P&P activities, one way to mitigate the risks is to break the P&Ps into three bundles to be done sequentially (if there is only one team). In this way, the committee has interim due dates, rather than one due date and should make progress throughout the allotted time. As the project progresses, risks that have been identified but avoided should be closed. The log should be updated and include the closed date. Risks that become issues should also be closed and include the date the risks became an issue, as well as the subsequent issue number. Any additional actions related to the risks would now be managed within the issues log.

Managing Issues

Unlike risks, the impact of an issue is happening now and needs to managed. Some issues from our case study could include:

- Change in the Tower opening date
- Loss of a resource to another project, employer, or due to illness
- Delay in hardware delivery
- Identification of five more ADT P&Ps impacted by the move

- Increased census requiring clinical staff (team members or stakeholders) to be more focused on patient care issues and not available for project work
- Additional software needed to develop specific training materials

Each of these issues needs to be addressed against its impact to the project. Issues can derail of project, and the PM needs to monitor for and manage any issues that develop. As with risks, once an issue is identified, the PM needs to add it to the issues log. Logs can be simple word processing or spreadsheet tables or managed through the organization's issues tracking application. At a minimum, the issues log needs to include:

- An assigned issues number (may be called a ticket number)
- Description of the issue
- The date identified
- Potential action to resolve the issue
- Responsible team member or stakeholder

As issues are worked, additional information should be added to the log including:

- Issue priority
- Issue impact
- Consequences if the issue is not resolved
- Status of the issue (new, in progress, completed)
- Updates to the issue
- Needed resolution date
- Issue resolution action (s)
- Change request number (if needed)

In a shorter or less complex project or when there are few issues, the PM can address most of the issues. Adding an hour or two to a team member's day may be an easy resolution to an issue. At times, the PM can do the task themselves to resolve the issue. Other times, the issue can be resolved by moving work to a future project or phase. In longer or more complex projects, however, issues need to be closely monitored and resolved as soon as feasible. If the Tower opening date is moved sooner, for example, the project timeline is negatively impacted and needs to be completely re-done. The PM needs to look at the options to either add more resources (either more people or longer working hours) or reduce the scope in order to complete the project in the new timeline.

Projects cannot always absorb more people in order to complete a task or resolve an issue. The added resources will need to be brought up to speed on the project and may adversely impact other work, either within this project or others. PMs often use the adage that "you can't have 9 babies in one month" to explain this. Certain tasks take a prescribed amount of time and cannot be completed sooner just by adding more people. For example, if it takes 3 months to order hardware and the Tower opening is moved earlier, it is likely that the hardware will not be delivered on time. While adding more people to the task will not help, in this case it might be possible to add an expedite fee to get the hardware sooner. The PM can add this to a potential action plan, but it will adversely impact the budget. Another way to manage an issue is to reduce scope – either removing tasks altogether or reducing the work to be done. It is unlikely, however, that tasks can be completely removed from a project without impact to the scope or project quality, or they probably would not have been included to begin with. The task scope, however, could potentially be reduced. Using the example of the Tower opening sooner, if the hardware cannot be expedited, then the PM can work with the team and the stakeholders to determine if existing hardware can be used instead or the new hardware could be installed after the opening date.

Using the training materials example, some of the training could be paper or lecture instead of the planned electronic or computer based training. Adding more classrooms, more teachers, or increasing the number of students per class could also help the training get done in a faster time frame. While moving the Tower opening date has obvious impacts, a delay in the opening could also adversely impact the project. If the Tower opening is delayed, the PM will need to monitor for the work filling the time or team members getting pulled away for other work and not available for the project any longer. While it might be seen as a good problem, this delay needs to be managed just like any other issue and the plan should be shared with the stakeholders in order to resolve the issue.

The PM continues to manage the work and control issues, risks, and changes until after all of the tasks are completed, the scope has been met and the project is live. Once this happens, the project moves into the closing phase.

Chapter 7

Closing – Novice

This chapter will cover the Closing process group related to novice project managers (PMs) and organizations without formal or mature processes. For those new to project management, there may be the assumption that the project is over when the final product or service is delivered. Closing actually begins at this time but doesn't end until the project deliverables are accepted by the sponsor(s), documentation is completed and archived, and resources are released.

The closing activities defined in this chapter may be completed at the end of the project or, for larger projects, at the end of each project phase. During this process group, all project documentation should be finalized and archived. One of the primary documents created during this time is the completion document. During planning, the project scope or charter defines what is included in the project and how it will be completed. The completion document defines what was included and how they were completed. It is a gap analysis of sorts and provides clear documentation of what was done during the project, what was delivered and allows for the project sponsor(s) to accept the project, as documented, as complete.

Finalizing Project Documentation

Documentation to finalize during project closing is listed below. These should be archived at the end of the project so they are available for future projects. During project initiation and planning, historical documentation from previous projects provide valuable information on how to complete similar projects, how to resolve similar issues, how to manage similar risks, and so on.

- Issues – verify all issues are closed and the documentation is complete. If any remain to be resolved, make sure they are assigned to someone for ongoing resolution. The documentation should be updated once resolved
- Risks – as the project is ending, all risks should be closed and any necessary documentation completed
- Workplan, Work Breakdown Structure (WBS), Schedule – all tasks should be updated to reflect the work completed with dates, durations, and resources
- Communication plan – update the plan with what was actually done during the project. Remember to conduct any planned communication related to project closure
- Any other planning or project supporting documents – these should be updated to reflect what was actually completed and how
- Support Documents – the documentation provided to the support team, or help desk, related to the ongoing support of the project deliverables. These should be up to date with description of what the project produced and how to support it. Documentation would include known issues that could arise and how to resolve them and common questions from training and the appropriate responses

Completion Document

The completion document, as mentioned above, is a new document produced during closing. This defines what was accomplished, how it was accomplished, and key information about the project. Once completed, it is signed off by the sponsor(s), which shows their acceptance and approval to end the project. The contents of the completion document include the following information, at a minimum:

- Scope Statements – list out the original, approved, scope statements, plus any approved scope changes. Each one should be identified if it was met or not; if not, a justification should be noted

■ Measures of Success – list out the original measures of success along with if they were met or not. The PM may include the actual measurements or details of how each was met

■ Assumptions and Constraints – note if these were accurate and how they may have impacted the project

■ Milestones – list the milestones from the original workplan, with planned and actual dates. Comments could be added to define any changes in these dates

■ Deliverables – identify the final deliverables with a brief description of each

■ Approvals – signatures and dates, these can be manual or digital signatures

There are many reasons for the information above to change between planning and closing that would impact how the project was actually completed compared to what was planned. Documenting what changed and how it impacted the project provides key information to help improve planning for the next project.

For example, the case study scope statements documented for the completion document could be as below:

■ Hire five experienced staff nurses
  – Not Met, only four nurses have been hired to date; the final position has been posted on external facing sites and interviews should occur later this month
■ Identify locations for 20 new PCs on the unit
  – Met, as of the project closing, only 18 have been installed; the final 2 will be installed within the next 2 weeks
■ Review all policies related to admission on the unit
  – Met
■ Develop test scripts for admission, discharge, and transfer of patients
  – Met

Lessons Learned

An additional activity to be completed during project closing is to conduct a Lessons Learned meeting. Everything that happens during a project is an opportunity to learn and improve for next time. Traditionally, this activity occurs during closing, but more PMs are starting to collect these lessons throughout the project to ensure nothing is forgotten and learning can begin right away. Since the PM is part of the project team, it is best if someone else facilitates this meeting, maybe another PM. This allows the PM to participate along with the rest of the team.

The Lessons Learned meeting should include all project team members. The purpose is to learn for future projects, and these lessons become historical documentation to be reviewed during future project initiation and planning. Since these are lessons, they should be documented as such. A comment stating, "I didn't know when my tasks were scheduled so I was always late," should be reworded to note what should occur in the future, stating the lesson. This could be reworded to say, "The PM should communicate the schedule to the team and provide reminders of upcoming tasks in each status meeting. This ensures all team members are available for their tasks when scheduled." The updated statement reflects the concept mentioned but is written to state what should happen and why. Noting the 'why' is as important as stating the lesson since it puts everything into context especially if the person reading this doesn't know anything about the actual project.

Once the completion document is approved, all other documentation is complete and archived, and the lessons learned are documented, the project resources can be released. They are now available to work on other projects or other assignments. This is the last activity, or task, in the Closing process group and officially ends the project.

This chapter reviewed some of the activities that occurred to close out a project. Ensuring complete documentation that is archived in a location where it can be used for historical/reference materials provides key information to help continuously improve the project management processes and improve planning of future projects. Before releasing the project team, remember to celebrate the project success.

Chapter 8

Initiation – Expert

This chapter will cover the Initiation process group related to experienced project managers and organizations with formal and mature processes. As opposed to Chapter 4 (Initiation – Novice), in these organizations, many of the activities discussed in this chapter have defined processes and procedures in place. Initiation begins with the request for a new project and ends when the project is authorized and a decision is made regarding when it will begin.

Project Requests

Chapter 4 (Initiation – Novice) defines a number of formal and informal ways a new request can be submitted. With formal processes and procedures, the formal submission process should be the norm with the informal ways discouraged. The defined procedure outlines what information should be provided in the request and often a template is utilized and is available to anyone who is authorized to submit new requests. The requestor rarely knows a lot of details beyond what they need. The request form can be simplified to gather key information from the requestor, such as listed below:

- Define the project being requested
  - What are the boundaries, or scope, of the request?
  - Is this related to a current application? If so, which one?
  - What is the expected timeframe for starting this project?
  - Is there a deadline for when the final product is needed?

- Define the business need
  - What is the justification for this project?
  - Is this related to a mandate or regulatory requirement?
  - Define the strategic alignment to the organizational or departmental goals?
- Define the funding for this project
  - What are the estimated project costs?
  - What are the estimated annual maintenance costs?
  - Are these costs included in a current budget. If so, which one?
  - When will the funds be available?

The process defines the procedure for submitting the request. It could be submitted to a person or a distribution group by email or directly into an application that provides an alert for the new request. Once received, it should be assigned for analysis and the gathering of additional details. This may be a member of the information technology (IT) project management office (PMO), but may not be the project manager who will be assigned to the project, if authorized.

A request could be submitted from the IT department for new projects related to operations and maintenance. These may include application updates provided by the vendor to fix identified issues, application version upgrades, data center hardware refreshes, database version upgrades, or operating system version upgrades. For these requests, the submitter can provide the information above as well as being the subject matter expert (SME) for information gathered in the request analysis.

Request Analysis

Once the request is assigned for the analysis, the goal is to gather enough information to determine what options are available to meet the need. The deliverable at the end of this step may be a change request (CR) if the request is smaller than the defined threshold for a project, a business case if the solution is not known and provides options, or a project charter if the solution is known and provides the details of the necessary work to meet the need. These are defined further in Chapter 4 (Initiation – Novice) as well as the typical information included in a business case.

With the deliverables in mind, the analysis includes gathering information from the following resources. The person completing the analysis should be knowledgeable of what questions to ask to gather the necessary details for the final documentation.

- Past projects – is there any historical information from past projects that would provide information on this request? The information could include stakeholders, assumptions, constraints, risks and how they were mitigated, tasks and activities, required resources, estimated durations, and any issues along with how they were resolved. Other project managers may also be a good resource for this information
- Vendors – if a vendor is known, what information do they have on how the request can be completed? This may include what work is required as well as any requirements for hardware or third-party software that may need to be purchased
- SMEs – these resources may come from anywhere in the organization depending on the specific request. They would provide their expertise to define what is required to meet the requested need, or what options are available

Project Charter

As mentioned above, Chapter 4 (Initiation – Novice) provides additional details on the information within a business case. If a project charter is appropriate for the request, there are more specific details related to the project needs for completing the requested need. The project charter typically includes the following information:

- Request/Project name
- Version number and date
- Project request information
  - Submission date
  - Submitter and contact information
  - Business owner
  - Desired start date
- High-level scope and requirements

- ■ Business impact
  - – Business functions/processes impacted
  - – Plans for ongoing operations
  - – Justification and how it fits with organizational goals and objectives
- ■ Resources requires, their level of involvement, and roles/responsibilities
- ■ High-level timeline with known milestones
- ■ Initial assumptions and constraints
- ■ Initial risks
- ■ Identified stakeholders
- ■ Security and privacy considerations
- ■ Success factors, or how will success be measured
- ■ Expected budget with high-level cost details
- ■ Approvals
- ■ Appendix for references or additional documentation to support contents

Governance

Once the analysis is completed and documented, there is at least one review of the request to determine the next steps. A formal process includes a governance committee who reviews requests from across the organization to make a decision on which projects should proceed based on the organizational needs, not the requestor's priority. The governance committee's charter will define which requests they will review and provide authorization to proceed versus those that do not require their input. This typically includes requests with any new applications or those above defined thresholds for work effort, duration, or cost. They typically do not include requests that are required or mandated based on security, privacy, or regulatory agency, general operations and maintenance, or those under the above thresholds.

In some organizations, the project request, along with the business case or project charter, is provided to the governance committee members prior to the scheduled meeting. This allows them to review the documentation and prepare any questions they may have for the requestor or the analyst. In some cases, the members are also asked to score or rate the requests based on defined criteria. The criteria may be related to how the request fits with the organizational goals and objectives, produces a benefit, provides operational efficiencies, improves the quality of patient care, or improves patient satisfaction. There is a total score based on simple addition or through weighted scores if some criteria are more important than others.

When the governance committee meets, they should be provided a status of the current project load and resource availability. This can be accomplished through a list of all projects in-progress, recently completed, and pending, with the associated resource requirements. This provides communication on all projects irrelevant to whether they reviewed them or not, as well as the resource capacity and demand to show the availability for the new projects. Once the committee understands the current state of projects and resources, they will review each new request. The request documentation is available for reference while the requestor provides a brief overview of the request. The members can then ask any questions about the request or the analysis documentation. Once all requests are reviewed, the committee discusses the requests and has the opportunity to adjust their scores or ratings based on any new information. At the end of the discussion, the authorization decisions are made.

The governance committee's charter should define if they have the authority to actually approve a request or if their decisions are recommendations. Typically, someone in the C-suite makes the final decision so the authority of the committee may be based on membership. As defined in Chapter 4 (Initiation – Novice), the final decision could be approved for further analysis of options, approved to begin now, approved to begin at a later date, or denied. The requests approved to start at a later date are often not the highest priority or may need to wait for available resources prior to starting.

Communication to the requestor throughout the process is important and should include when the final decision is made and the expected start timeframe. All documentation should be stored in a location available to others who will be working on this request when the project begins. The Initiation process group ends with the final authorization decision, even if there is a delay in the project beginning.

A simplified request and governance process is shown in Figure 8.1.

The Initiation process group begins with the request and ends with an authorization decision. Since the project manager, or team, is not often involved in this activity, it may appear to be skipped. Chapter 9 (Planning – Expert) describes the Planning process group. When the project is ready to begin and the project manager is assigned, planning will begin. Planning takes the information gathered here and further defines the project boundaries and the work required to complete the final deliverables. Chapter 9 (Planning – Expert) discusses project planning for the more expert project manager.

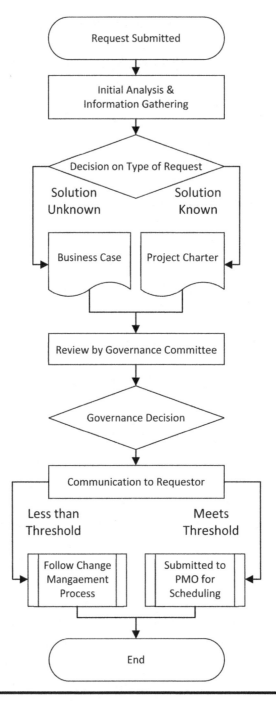

Figure 8.1 Request and governance process.

Chapter 9

Planning – Expert

As we have discussed in Chapter 5 (Planning – Novice), detailed planning increases the likelihood of project success. Having scope, resource, communication, change management, and quality management plans will help the project manager (PM) execute and control the project. Planning documents drive the work, how the PM will control the work, and reduce the potential for changes to the project scope. Chapter 5 (Planning – Novice) described basic scope, resource, workplan, and communication plans. Longer, more complex projects require more detailed plans and some additional tools.

Developing the Workplan

As discussed in Chapter 5 (Planning – Novice), creating the workplan is an art and a science. Identifying all the tasks needed to complete the approved scope, along with the duration and responsible person for each task, takes time and focus. Experienced PMs often use workplans from previous projects to help create new workplans, as the basic structure is usually the same. All projects include some form of analysis (plan), build, test, train, go-live, and closure phase. Starting with each of these phases helps the PM organize the tasks into building blocks that can be developed through the planning process. Using the training example, tasks need to include, at a minimum, identifying:

- What will be included in the training (e.g., new policies)
- How the training will be delivered (e.g., computer-based)

- Who will do the training (e.g., existing trainers)
- Who needs to be trained (e.g., all clinical staff assigned to units in the new Tower)
- Where the training will take place (e.g. via computers on the units and accessible from home)
- When the training will occur (e.g., 2 weeks before go-live)
- The actual activities to create the training

Most PMs add tasks to the workplan in the order in which they will occur. Working with the stakeholders and team members to determine how long an activity has taken in the past will help identify current project task durations. Developing the workplan is an iterative activity. Often the durations are modified as the workplan is created, as there is often a pre-defined go-live date. For example, if, in order to meet a pre-defined go-live, the task durations need to be reduced, the PM may need to then adjust scope (e.g., reduce the number or duration of the computer-based training activities) or add additional resources in order to successfully complete the tasks. While there is no hard and fast rule, PMs should try to keep tasks to no more than a 2-week duration when possible. Tasks can shorter than 2 weeks, but breaking bigger tasks down can make it easier to both get the work done and track the work accordingly. Breaking down bigger tasks into 2-week chunks of time can help keep the resources focused on completing the tasks on time. Frequently, team members start tasks on time but may not attempt to complete it as soon as possible, especially if they have a long time to work on the task. By keeping each individual task to a 2-week duration, the work will be progress and is more likely to be completed as scheduled.

As tasks are added to the workplan, the PM will assign resources to each task. If the PM is using a project management application, there are two ways to assign resources in the plan. The PM can add resources to each task and allow the project management application to fill the resource time with the duration of the task. For example, if a task is 2 weeks long, the resource will be assigned to 80 hours of work, but default. The PM, however, can modify the amount of time a resource spends on a task. For example, the task may be spread over 2 weeks, but the resource may only spend 8 hours a week on that particular task. The PM can adjust the work hours for each task so that the task continues to be a 2-week task, but the resource only has 16 hours on that task. Each task should be managed this way in order to best estimate the total amount of time resources will spend on the project. Often when tasks are assigned to a project full time, PMs do not track the actual time resources spent on the project. This, however, prevents future projects from being accurately estimated. Instead, the actual amount of time resources is expected to spend on the task should be added to the workplan. Once all the resources and tasks are added, the PM can level the resources. This allows the PM to see where tasks may not be completed on time due to resource constraints.

As tasks are added, they can also be identified as milestones. A milestone task is a major project goal or deliverable. It is often a task that is required to move on to the next part or phase of the project. Example milestones in our case study could be:

■ Training materials completed
■ Computer equipment ordered
■ Take delivery of computer equipment
■ Testing completed
■ All room and bed updates in electronic medical record (EMR) completed

Questions that help the PM determine which tasks are milestones include determining if the task:

■ Is a deliverable
■ Will directly impact the go-live date
■ Will prevent the project from moving to the next phase
■ Needs sign-off from the sponsor(s) before moving to the next phase or tasks

Milestones can also be activities outside of the project that are needed to move forward with this project. For our case study, this could be major construction activities, such as completing the networking needed for the computers or finishing the interior work on the units. Until these tasks are done, this project cannot progress with installation in the new Tower. Milestone tasks can be added to a workplan as a task with no duration and marked as such in project management software. Milestones are very useful when executing the project workplan, as they quickly identify major activities that must be completed on time in order to keep the project on schedule.

Another milestone the PM can add to the workplan are go/no-go decision points. Throughout the project, there are times when moving to the next phase is completely dependent on prior phase work being completed. For example, moving to the testing phase before the EMR software has been revised or the hardware prepared will probably lead to unnecessary testing errors. Adding the go/no-go milestone does not increase the overall workload but can help show readiness to move to the next steps. The PM can work with the stakeholders and sponsor(s) to define the go/no-go criteria, as well as who will approve the 'go'. See Table 9.1 for an example of go/no-go criteria for our case study.

In addition to adding milestones to the workplan, experienced PMs often also add dependencies to the tasks. Adding dependencies to tasks helps to determine the time needed to complete the project. For example, if determining who will review each policy and procedure (P&P) has to be done before the P&P review can start, then the task to review the P&Ps is dependent on the task to determine who will review the P&Ps. There are two

Table 9.1   Example of Go/No-Go Criteria

Go/No-Go Criteria	
Communications Plan	Includes pre-live and post-live communication (audience, message, vehicle, etc.)
Policies and Procedures (P&Ps)	All affected P&Ps have been revised and published
Training Complete	Goal is 100%
Build Complete	Goal is 100%
Testing Complete	Goal is Zero Critical Software Issues
Hardware Tested	Hardware testing includes testing of wireless connectivity on the units deploying the hardware, 'fitting' into the point-of-care location, access to production software, and access to printers for demand printing
Hardware Deployed	Goal is 100% deployed by go-live
Support Schedule Complete	Includes support staff schedules, executive staff, and super user schedules for LIVE period
Issues Management Plan Complete	Includes how to collect, resolve, and report issues

types of dependencies that can be added to the workplan: predecessors and successors. Predecessors are activities that must be completed or at least begin before other tasks. A predecessor task controls the start or end date of one or more other tasks. Successors, on the other hand, follow other tasks. Successor tasks can follow the start or end of another task or begin after some identified duration. Dependencies can be established as start-start (both tasks need to start at the same time), start-finish (the successor task cannot start until the predecessor task is finished) or finish-finish (both tasks need to finish at the same time). Lag time can also be added to the task dependencies. For example, developing the training materials is a successor to P&P revisions. The PM can add some lag time for the trainers to gather any additional information needed before they start developing the training materials. The development training materials task is a start-to-start task with lag. See Figure 10.2 in Chapter 10 (Execution, Monitoring and Control – Expert) to see how dependencies are viewable in project management software.

The PM continues to add tasks, durations, resources, milestones, and dependencies until all the identified activities are included in the workplan. Once the workplan is completed, the PM should obtain sign-off and baseline the project. The initial baseline formally defines the approved project workplan to include the approved scope, resource, and schedule. Once a project is baselined, it can be used to measure how project performance deviates from the plan if any changes occur once the Execution phase starts.

Stakeholder Management Plan

PMs need to manage resources in a project, including stakeholders. Stakeholder management is about the politics of any project. Stakeholders often have different goals and objectives that could lead to differing expectations and requirements of any project. Stakeholder management is a strategic activity that PMs use to maintain support for the project. A stakeholder management plan is used to determine the appropriate methods to effectively engage stakeholders throughout the entire project lifecycle. It helps the PM to understand each stakeholder's needs, interests, and potential impact on the project success. This helps the PM to better communicate with the stakeholders in a way that resonates with them. The plan starts with identifying the people and/or departments that could affect or be affected by the project. For the new Tower project, this is almost all the departments in the organization. For individual projects related to the Tower opening, however, not all departments may be directly impacted. For example, deploying new hardware in the new Tower units impacts or is impacted by physicians, unit staff, clinicians, nurses, information technology (IT) resources, and potentially housekeeping if they are responsible for cleaning the equipment. It is unlikely departments such as radiology would be stakeholders in the hardware project as they will not use the new hardware, but instead use hardware they bring with them (i.e., X-ray machines) or the hardware already in their departments.

Once the stakeholders are identified, the next step is to analyze stakeholder expectations and their impact on the project. Many clinical users see new projects as a way to resolve existing issues. Using the hardware project as an example, if physicians, nurses, and/or clinicians do not like the existing hardware or want more expensive options, it is imperative to understand their expectations of the project in order to manage them. Once stakeholder expectations are identified, the next step in the stakeholder management plan is to develop appropriate strategies and tactics for effectively engaging stakeholders in a manner appropriate to the stakeholders' interest and involvement in the project. Nursing and clinical department stakeholders will want more information about the project and will often want it sooner than typical physicians who usually want the information only when they need it. Both types of stakeholders' expectations need to be managed in order to keep them engaged in the project. Creating a stakeholder management grid and following it throughout the project will help the PM to successfully manage stakeholders. At a minimum, the stakeholder management grid should include:

- Stakeholder Department
- Impact on Project
- Impacted by Project
- Current State
- Desired State
- Issues, Opportunities and Risks
- Mitigation Strategies and Actions

Once this information has been gathered, the PM can create a stakeholder analysis matrix and locate each stakeholder in the appropriate square showing the intersection of their impact and influence. Each stakeholder should be placed in the square that best describes their influence on and impact from the project. This will help the PM to determine how to

Table 9.2   Stakeholder Analysis Matrix

Impact ↑	Passive (keep satisfied)	Key player (manage closely)
	Distant (monitor)	Important (leverage their influence)
	Influence →	

communicate with each stakeholder. See Table 9.2 for an example of a stakeholder analysis matrix.

Another tool used to manage stakeholders is the RACI diagram. RACI stands for responsible, accountable, consulted, and informed, and describes the way the PM should plan to communicate with each stakeholder and the sponsor(s) about each activity in the project. The RACI diagram helps with the communication structure and further clarifies the role each stakeholder has within the project. It also helps to ensure that all the major activities have someone assigned to be responsible and accountable for them. The four roles that team members and stakeholders can play in the project are:

■ Responsible – team members or stakeholders who must complete the activity or make the decision. There can be several roles in the responsible column

■ Accountable – the owner of the activity who must sign off once the activity or decision is completed. There should be only one accountable person per activity or decision

■ Consulted – team members or stakeholders who may need to give input before the activity or decision is signed-off. These roles are often active in the project, but not decision makers. There can be several roles in the consulted column, but the number should be kept to a minimum

■ Informed – team members or stakeholders who need updates on progress or decisions but do not contribute directly to the activity or decision. There can be several roles in the informed column, but the just as with the consulted column, the number should be kept to a minimum

The PM creates the RACI diagram by listing all of the major activities and known decisions in the left column of the diagram and the stakeholders in columns across the top. One column for team members is often enough,

Table 9.3 Example of RACI Diagram

	Sponsor	Project Manager	P&P Committee Chair	Training Lead
Determine which P&Ps need to be reviewed	A/R	C		I
Determine who will review each P&P	C	I	A/R	
Communicate due dates to reviewers	C	A/R	C	I
Review P&Ps		C	A/R	I
Approve revised P&Ps	A	C	R	I
Develop training (if needed)	I	I	C	A/R
Deliver training (if needed)	A	I	I	R
Publish new P&Ps (go-live)	I	I	A/R	I

but each team member could be listed individually if needed. Then the PM identifies the responsible, accountable, consulted, and informed stakeholder for each activity. It is important to make sure that every task has at least one stakeholder in the responsible column and no more than one stakeholder in the accountable column. See Table 9.3 for an example of a RACI diagram.

While this can be a time-consuming activity in large project, identifying the responsible, accountable, consulted, and informed stakeholder(s) for each activity will help the project run more smoothly.

Risk Management Plan

Risk management is the process of identifying, assessing, responding to, monitoring, and reporting risks. In larger or more complex projects, the PM needs to create a risk management plan. It defines how risks will be identified, analyzed for potential impacts, managed to mitigate the risk, and recorded throughout the project. Once created, this plan should continue to be reviewed and updated for the rest of the project. Once the risks are identified, the PM can create a qualitative risk analysis grid with probabilities and impacts of the risk if it becomes real. The grid includes high, medium,

Table 9.4   Example of Qualitative Risk Analysis

		Low (below 30% probability of occurrence)	Medium (between 30% and 69% probability of occurrence)	High (greater than 70% probability of occurrence)
Impact	**High** (potential to greatly impact project cost, project schedule, or performance)	#3		#4
	Medium (potential to slightly impact project cost, project schedule, or performance)		#6	#2
	Low (relatively little impact on cost, schedule, or performance)	#1	#5	
		Probability		

and low probability and impact. Risks are placed in each box depending on their probability of occurring and the impact to the project if they do occur. Risks at the top right of the grid hold the highest probability and impact, are considered major risks, and need to be closely monitored. See Table 9.4 for an example qualitative risk analysis grid.

For each major risk, one of the following approaches can be used to address it:

■ Avoid – eliminate the risk by eliminating the cause
■ Mitigate – identify ways to reduce the probability or the impact of the risk
■ Accept – do nothing
■ Transfer – make another party responsible for the risk

Different organizations handle risk differently, so the PM needs to work with the sponsor(s) to identify the correct strategy for each risk that would be acceptable in this project. Once identified, the approach should be added to the risk log and followed throughout the project.

Issues Management Plan

The PM also needs to develop an issues management plan that describes how issues will be managed, prioritized, escalated, and controlled throughout the project. Similar to the risk management plan, the issues management plan is the process of identifying, analyzing, resolving, monitoring, and reporting issues. The goal of issues management is to prevent issues from negatively impacting the project. The issues management plan should include at a minimum:

- How issues will be documented (standard issues template)
- How issues will be managed (logged and reported)
- How issues will be prioritized
- How issues will be analyzed
- Who the decision makers are

The issues management plan should define a streamlined and simplified issue escalation and resolution processes to resolve issues quickly and efficiently. Prioritizing issues helps to concentrate the work effort on issues that have the greatest impact on the schedule and quality. This will help the PM focus the sponsor(s) and stakeholders on the key issues that could negatively impact the project. See Table 9.5 for an example of issues priority descriptions.

Setting a predefined criteria and process to determine initial issue ownership and ensuring participants have clear project roles and areas of accountability help to enable clear, quick issue escalation and resolution. The roles and responsibilities for issues management should be clearly spelled out in the issues management plan. See Table 9.6 for an example of an issue management plan roles and responsibilities grid.

Once identified, issues should have a resolution date and back-up plan if the issue cannot be resolved by that date. This expectation should be clearly spelled out in the issues management plan. The issues management plan should also spell out how planned resolutions and target dates will be communicated to all team members and stakeholders. The issues log should be available to all team members and stakeholders for review as needed. Status reports should include information about the high impact issues to limit surprises throughout the project. Issues will happen in any project, so having a clear, organized plan will help the PM manage them throughout the project. In addition, keeping the stakeholders apprised of issues and how they are being addressed can help to minimize their concerns.

Table 9.5 Example of Priority Descriptions for Issues

Priority Level	Prioritization Guidance
Critical	• An issue that will affect the go-live timeline • Any patient safety issue
High	• Any issue that has a negative impact on patient care • Any issue that does not have a "work-around" (business or clinical process) • An issue that has a high financial impact • An issue that has a high impact on workflow
Medium	• An issue that impacts a specific area of testing, but activities can continue in other areas or a 'work-around' for the issue is available until the issue is resolved • An issue that impacts a business or clinical process, but a 'work-around' is available until issue is resolved • A system or business issue without a reasonable "work-around" but does not impact patient care • An issue that has a low financial impact
Low	• An issue that impacts a small number of users • A request that is cosmetic in nature • An issue that has no financial impact

Change Management Plan

Projects will include requests for changes to scope, budget, and/or timeline. As part of the Planning phase, the PM will create a change management plan to define how change requests will be managed throughout the project. The purpose of a change management plan is to control change during the project and describe the roles and processes need to control change. Change is defined as anything beyond the original scope, budget, timeline, or resources once scope sign-off has been obtained and the project is baselined. The change management plan should include the following four components:

1. roles in the change management process
2. final decision makers
3. process to request changes
4. standard change request form and log

Table 9.6 Example of Issues Management Roles and Responsibilities Grid

Role	Responsibility
Issue Originator	• Documents the issue as clearly and completely as possible using the provided template • Re-creates issue to validate system issue and not user error • Submits issue to project manager/issues log
Project Manager	• Tracks status of issue in the issue log • Updates issue log and assigns priority • Assigns issue for impact analysis • Presents recommendations to sponsor(s)
Analyst (team member)	• Researches and clarifies issue as needed • Identifies alternative resolutions • Makes recommendation • Estimates time and resources required to resolve the issue • Identifies where added tasks fit into project plan • Updates issue form with impact analysis • Sends updated issue form to project manager • Updates functional and application specifications as necessary
Sponsor(s)	• Either approves the resolution, denies the resolution, or places the resolution on hold • Ensures any required additional resources are available

Roles in the change request process should include at a minimum, the PM, sponsor(s), stakeholders, team members (who will complete the analysis of the change), and the final decision maker(s). Often, it is the PM or a team member who does the formal request, but anyone who is involved in the project can request a change.

In some projects, the sponsor(s) is the final decision maker, but in larger, more complex projects, a change control board is convened. The change control board should include a stakeholder from all of the departments or areas impacted by the project. While there may be more than one stakeholder per department involved in the project, only one should be a member of the change control board. This way, all departments or areas are represented equally. The PM should be a member of the board to represent the overall project. The project sponsor(s) may be the chair of the change control board or receive the recommendations from the board and approve the changes. The change control board should meet regularly so that change requests can be assessed frequently. The final change request decision maker is usually defined by the organization's culture. Some organizations allow the stakeholders to make decisions as a group. Others allow the stakeholders to make recommendations, and the sponsor(s) is the final decision maker. Still others allow a certain level of change decisions to be made by the stakeholders, while others change decisions need to be escalated to a higher authority. For example, change requests that do not adversely impact cost, schedule, or resources, that is, changes that can be absorbed within the project, could be approved by the stakeholders. Change requests that require a change to the budget, major milestones, such as start of training or the go-live, or resources, however, could require sponsor(s) approval.

Once decision makers are defined, the PM needs to develop the change request process. The process for submitting, analyzing impact, and approving the change should also be part of the change management plan. All changes should be documented via a standard change request template. A sample change log should also be included as part of the change management plan. At a minimum, the change request template should include the following information:

- Date change request created
- Change request number (usually assigned by the PM)
- Description of the requested change
- Name of person submitting the change
- Anticipated benefit to department or area of the change
- Anticipated impact to department or area of the change is not implemented
- Impact to project of doing the change (to time, budget and resources)
- Recommendation (to do or not do the change)
- Change control board and sponsor(s) signatures
- Date change approved
- Status (such as open, work in progress, in review, testing, closed, not approved)

The change request process should include the following steps:

1. All change requests should be logged in the change request log
2. Once entered, a change request number is assigned by the PM

3. The PM assigned a team member to analyze the request
4. The request is presented at the change control meeting
5. Change control board approves requests additional information or denies the change requested
6. The change is added to the project workplan
7. The change log is updated with each step

Once all of this information is added to the change management plan, the plan is ready for approval.

Quality Management Plan

Managing the quality of a project is crucial to the overall project outcomes. Project success is often measured by the stakeholders as 'useful', rather than complete. If the project does not include quality deliverables, the project is often not considered a success. One way to increase the likelihood of good-quality deliverables in a project is to create a quality management plan during the Planning phase. The purpose of the quality management plan is to describe how quality will be defined and managed throughout the project. Quality management plans are most frequently used in software projects but can be created for any type of project. In our case study, modifying the EMR to accommodate the new units, rooms, and beds in the Tower would be a good use of the quality management plan. In addition to defining how quality will be defined and measured the quality management plan also defines what standards will be followed, and the roles and responsibilities of the team members and/or stakeholders involved in defining and measuring quality. In larger, more complex projects, the project itself is held to standards of quality, which are often assessed through a project audit. The project audit looks at the quality of project documents such as issues and change control logs, the workplan, status reports, and risk mitigation plans. These audits can take place before major milestones, such as the Planning phase, to make sure that project quality is defined and can be followed throughout the project.

Deliverable quality is also defined in the quality management plan. The plan starts with a definition of the project assumptions and constraints. For example, using our EMR modification project, assumptions would describe all of the tables that need to be updated with the new units, locations and service lines. Next, the standards to be followed would be defined. For example, final testing should have zero errors and will follow the organizations standard IT test plan. Last, the plan would describe how the PM will be monitoring quality and how quality will be documented. For example, the test log will capture each test event and the success or failure of each step, as well as the overall success (as a percentage) of each event. Defining the plan to measure and maintain project quality during the Planning phase will help the PM keep the project on track.

Success Factors

Determining how project success will be measured is part of the project Planning phase. The project itself will be measured by metrics such as number of issues, changes to schedules, budget, and timeline. Using benchmarks from other projects will help the PM determine what should be expected from this project. PMI uses the term *project benefits management plan* to describe this plan. Measuring organizational success, however, uses different metrics based on the scope of the project. For example, since this new Tower project may have new P&Ps, organizational benchmarks could include things like:

■ Average time spent in the ED before patient is admitted to the hospital as an inpatient
■ Medication errors
■ Length of stay
■ Medication reconciliation completed upon admission
■ Average adjusted expenses per inpatient day

The organization benchmarks should be something the organization is already collecting so that they can see the impact the new Tower projects have on their organization.

Metrics and benchmarks are identified during the Planning phase. The PM will identify project-specific metrics to be monitored and create a plan to monitor throughout the project. Organization metrics are usually identified by the stakeholders. Often, the PM coordinates the activity, but stakeholders develop the details of which metrics will be assessed as part of the project, how data will be collected, when data will be collected, and what will define success.

Pre-project data collection can start as soon as the plan is completed.

Test Plan

Another plan that is completed during the Planning phase is the test plan. The test plan outlines how the testing phase of the project will be carried out. It defines what will be tested, how it will be tested, who will do the testing, and when testing will occur. There are a variety of testing types, and not all are appropriate for every project. Many people are familiar various testing types, such as unit testing, integration testing, security testing, regression testing, validation testing, parallel testing, and user acceptance testing. Unit testing, integration testing, and regression testing may appropriate for our case study example of updating the EMR to include the new unit and beds. Projects that are not related to software, such as the case study for developing the new P&Ps still require testing. For this type of project, testing is usually referred to as validation or verification testing, since the documentation is not actually being tested. In this example, the testing process is to ensure it that the P&Ps are complete, include the required information, and are formatted correctly.

The test plan also defines how the testing will be completed. It defines whether testing will be completed manually with detailed test scripts or testing scenarios or automated with a testing tool, or in the case of validation, reviewed by a committee or stakeholders. If the testing is for software, the test plan needs to define which environment will be used; typically, there is a test environment with test data that should be available. Whatever type of testing is required for the project, the test plan should include the following information, at a minimum:

- Project information
- Testing resources
- List of what is being tested/validated
- List of what is not being tested/validated
- Type of testing to be completed
- The approach for testing (how the testing will be completed)

- Estimated timeline for all testing activities
- Testing deliverables
- Pass/fail criteria

In addition to the above items, the PM needs to determine who will sign off on the actual testing as part of the plan. Sign-off could be assumed if all testing tasks pass, but if they do not, the test plan needs to define the process of repeating the testing if needed. At times, a less than 100% pass rate is acceptable, as long as there is a plan to correct the items that failed the test. For example, if 95% of the hardware works, the clinical staff in the new Tower can most likely find a working device. If, on the other hand, only 95% of the room and beds are correct, critical patient care could be delayed and there could be bad outcomes. Either way, the criteria should be defined as part of the plan. Once the testing plan is approved, the testing tasks should be added to the workplan.

Activation Plan

The activation plan defines what will occur surrounding the go-live. The plan should outline what the activities are for each phase of the go-live. Go-live phases include all preparation activities, the actual activation, and the follow-up activities until everything within the scope is live (often called post go-live). This is a key point if the project will be implemented in phases or one unit at a time as there may be multiple activation periods.

The plan should also include the roles and responsibilities required for the activation. These would include who has the authority to make the key decisions such as the date and time the activation will occur or if the activities should be rehearsed to identify any issues or problems prior to the activation day. There are many decisions that are required, and they vary across different projects. One critical role is the facilitator of the activation activities, which is typically the PM. The plan provides guidance as the activation planning occurs, which should begin as soon as possible. Most of the details around the actuation planning occur during the Execution and Control phase, but the plan should be developed during the Planning phase to create the guideposts for the activation. As with the testing plan, activation activities should be added to the workplan once the plan is approved.

Planning Phase Sign-Off

Once all the plans are developed, they require sponsor(s) sign-off. Presenting each plan as its ready will make the sign-off process more successful. The use of a standard sign-off form will also help make the process more efficient. The sign-off form should include, at a minimum:

- Name of plan
- Date submitted for approval
- Name and project role (i.e., sponsor(s), stakeholder, etc.) of approver
- Signature of approver
- Data approved

When the plan is ready for sign-off, it is good to save the plan as a PDF rather than an editable document in order to maintain change control of the documents. In this way the approved document will be maintained and any changes will follow the change request plan. Approval sign-offs should be maintained with the final version of the plan and stored with the project documents. Approval of the planning documents marks the end of the Planning phase.

Chapter 10

Execution, Monitoring and Control – Expert

As discussed in Chapter 6 (Execution, Monitoring and Control – Novice), much of the project manager's (PM) work during this phase is to ensure that the work gets done and the project remains on time, on budget, within scope, and meets the stakeholder's requirements. As in smaller, less complex projects, the PM must manage the plans created during the Planning phase and communicate project status with the stakeholders. Larger, more complex projects require even more diligence to manage the project and prevent scope creep. While the workplan, issue log, risk log, and status reporting are managed during this phase, experienced PMs will use additional or more complex tools to control the project.

Managing the Workplan

As described in Chapter 6 (Execution, Monitoring and Control – Novice), the PM monitors the project work against requirements and the workplan and should update the workplan regularly. The PM should update each task with status and the actual start and end date if they differ from the original date. In larger, more complex projects, most PMs use project management applications or tools to help them manage the workplan. Project management applications can help the PM to plan and organize tasks, develop estimates, manage resources allocation, and provide automated reports to stakeholders. In addition, many applications include methods to notify resources of their tasks, changes to their tasks, and allow resources to update their own tasks. There are many PM applications available; some are free; others require organization versions. Some of the free versions limit the duration or number of projects; others limit storage but can still be useful tools. Many come with free trials, but the free trial may not be long enough for a complex project that requires project management software. A spreadsheet application can be used to do basic task management but does not provide automated reports, nor does it automatically notify team members or stakeholders of changes. That said, however, some organizations only use spreadsheets for project management and are able to manage projects successfully. PC Magazine (pcmag.com), Project-Management.com, and others list and rate project management applications each year. These sites list the top applications along with prices, benefits, and issues to help select the application that will meet the needs of most PMs. Most organizations provide a standard application for PMs, so the best application is the one available to you when you are managing a project. It takes some time to get used to project management applications, as they come with defaults on how resources are managed as well as standard reports that can be modified for your project. Many tutorials are available for free either from YouTube or once you have access to the application.

One of the biggest benefits of project management applications is the many views they provide of the project. A Gantt chart shows the overall project schedule and dependency relationships between the activities. Often the Gantt chart view gives the most overall information about the project. Project management applications give much more detailed Gantt charts than spreadsheet applications. (The authors do not endorse any specific project management applications but have used examples from applications we have access to.) See Figures 10.1 and 10.2 for examples of workplan Gantt views in Microsoft Excel and Project, respectively.

Other automated views are available in project management applications to help the PM manage the workplan. These views include, but are not limited to, critical path, milestones, overdue tasks, tasks that have not

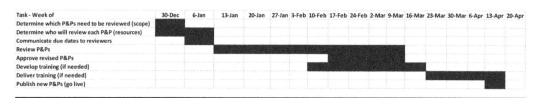

Figure 10.1   Example of Excel Gantt chart.

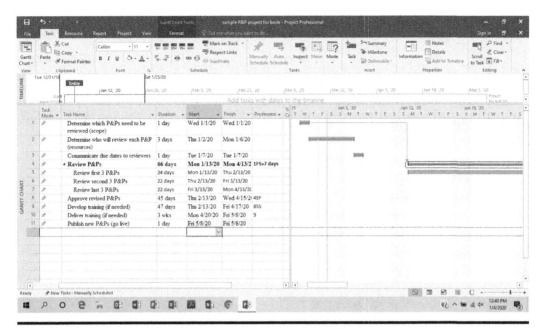

Figure 10.2   Example of Microsoft Project Gantt chart.

started, resources that are over allocated, and general project timelines. By looking at the workplan in a variety of ways, the PM can use these applications to get a better idea of where the workplan issues are throughout the project. No matter which application is issued, the PM's activities remain the same during this phase. Each task has a start and end date, assigned resource, and may have predecessors and successors. Each task should be updated as it gets started and as it progresses to completion. Team members should be able to give the PM updates on the percent complete of each task they are assigned, and the PM can monitor the workplan accordingly. For example, on day 2 of a 5-day task, the task should be 40% completed to remain on target. Tasks that have not started as scheduled or are not on target should be reviewed with the responsible resource to determine how to get the task back on track. If tasks are behind or have not started as scheduled, the PM may need to modify the workplan. Major changes to the plan should be followed by a re-baselining, so that the PM can understand where the project is and what impacts any delays may have. Predecessors and successors help when task dates need to be changed. For example, if the policies must be completed before training is completed, these tasks can be related to each other in the workplan with a finish-to-finish predecessor. In this way, if the review task gets delayed, it is easy to see the impact on the training tasks. The more

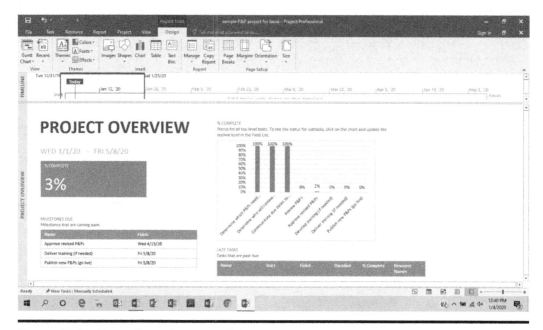

Figure 10.3 Example project overview report.

relationships that can be identified and added to the workplan, the easier it is to see the impact one change can have on other project work.

As the workplan gets updated during the project, there are also many reports that can be generated from project management applications that will help report to stakeholders. One report that is useful is an overall project status that includes how much work has been completed to date. See Figure 10.3 for an example project overview report.

Other available reports include critical tasks status, resource allocation, slipping tasks, milestones status, and burn-down, which shows the amount of work complete against that which is still to do. Some of these reports could be used during status meetings to show project progress. Using these views and reports will help the PM keep the project on target.

Managing Scope

As discussed in Chapter 5 (Planning – Novice), the PM needs to manage the triple constraint – scope, time, resources – to make sure they remain within the planned balance. Much of the effort during this phase is working on the balance of these constraints. At its basis, change in one of the triple constraints requires some change in one or both of the other constraints.

An imbalance can endanger the project's success. A lot of the PM's effort during this phase is focused on managing scope. The workplan is a list of tasks that were determined during the Planning phase. Completing these tasks according to the timeline and quality plan should be the main focus of the project during this phase. Stakeholders and team members may identify additional tasks or activities needed as they work through the project. Any requested changes need to follow the prescribed change request process in order to minimize scope creep. Longer, more complex projects are especially prone to scope creep due in part to their longer duration. Changing priorities, technology, and resources can greatly impact project success and add to scope creep.

There are a number of ways to manage scope creep. The first, but most unlikely to work, is to say no to every request without analysis. In a very short project, this may be an effective method as it won't be long until post-project activities can start, but most times it is not. The best way to manage requests is to treat each request the same and follow the change management process approved during the Planning phase. Even seemingly small or simple requests need to follow the process to maintain full control over the scope. For example, a request may seem simple and could be completed within an hour. One such request probably will not derail the project, especially if done early in the project. However, often one approved request encourages more requests that the team and/or stakeholders expect to be approved in the same way, especially if they are considered simple changes. One- or 2-hour-long requests can probably be easily absorbed into the project. It is a slippery slope, however, and requests can easily become an avalanche that will derail the project. If the project is 100 hours of work, 1 hour is 1% of the allotted time. Five extra hours of work, however, means the project is at risk of not meeting on time, on budget expectations, and something else may have to be forfeited.

Even a simple 1-hour change request can adversely impact the project if it is requested later in the project, since a lot of the work has already been completed. Each change needs to be tested in relation to the rest of the work and may require testing more than just the change itself, changes to the training materials, or impact other areas of the project. The later in the project, the more risk the change adds. Risks, in general, are reduced as the project lifecycle advances since the may have been avoided. The cost of a change, however, can increase as more of the work has been completed and may have to be re-done to meet the requested change. See Figure 10.4 for a graphic representation of the cost of changes.

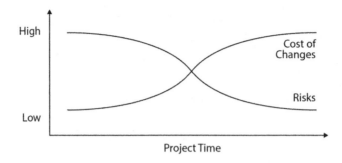

Figure 10.4   Costs of changes.

This is where the change management process comes in. The change management plan defines how changes will be managed. Requestors need to complete a formal change request for every change, no matter how small. In this way, all requests will be evaluated using the same process which will help reduce scope creep. Each change request should include, at a minimum:

■ Change request number
■ Description of the requested change
■ Description of why the request is necessary to the requestor (benefit)
■ Date requested
■ Potential impact if the change is not implemented

One the PM receives the request, it should be added to the change log and assigned to be evaluated. Evaluation should address:

■ Who will evaluate
■ Expected impact to the project scope
■ Expected impact to the project cost
■ Expected impact to the project schedule
■ Expected impact to the project resources

When assessing impact of a request, it is necessary to understand the value proposition. Changes need to bring value to the project in order to be worth the added effort, cost, time, and/or risk. One way to measure a change's value is to use a formula. For example, Value = Quality/Cost. While it is usually difficult to identify actual dollar amounts quickly, estimates can be used to help determine the impact of the change. For example, if the change request is to add 20 additional computer carts, the PM will need to determine the cost of the carts, any potential fees to expedite delivery, and additional time or resources needed to test and install the carts. For this project, we will assume each computer carts costs $4000, so the additional carts requested will cost $80,000. According to the requestor, the additional carts will save each nurse 5 minutes per shift since they will be able to have a cart at every bedside rather than having to move them around. This 'value' could also be estimated. For this project, we will assume 20 nurses per shift at $45.00 per hour, which comes out to about $4.00 per 5 minutes. There are three shifts, so the value of the time saved is about $1200.00. Quality from the additional carts will have to be estimated using items such as a potential reduction in errors or increased completeness of documentation. While quality is very hard to estimate, some dollar amounts could be interpolated from current error rates and costs, Medicare denials, length of stay, etc. Adding this information to the change request is the role of the PM with help from select stakeholders. Once all of the evaluation is complete, the change control committee can make a decision. Once the determination is made, it should be entered into the log. If the change is not approved, the change request should be closed or moved into a future status. If the change is approved, the PM needs to adjust resources, budget, and/or timeline to accommodate the change. In addition, the timeline may need to be re-baselined to account for the changes.

Quality Management

In addition to managing scope to avoid scope creep, the PM needs to manage the quality of the work. Large, complex projects often start with documented requirements. Using these requirements will help to manage for quality. In general, quality is defined by the stakeholders. They need to be satisfied that the project met or exceeded the requirements. Sign-off from the stakeholders throughout the project can help it remain on target. Quality sign-off should be completed as each phase or milestone is completed and maintained in the project folder. Often a simple form can be used to gather signatures. The form should include, at a minimum, a description of the item, task, or phase that is being addressed along with a space for the appropriate signatures. Sign-off could be collected for a single deliverable, assuming that the rest of the tasks in that phase are part of that deliverable or more of the same. For example, one computer cart could be tested and installed on an existing unit for stakeholders to review and sign off. Then, the rest of the carts can be readied following the same requirements.

Final sign-off would be when all the carts were installed in the new Tower but would not require detailed testing, just connectivity testing.

Issues and Risk Management

Another key activity for the PM throughout the Execution and Control phase is issue management. While basic issue tracking was addressed in Chapter 6 (Execution, Monitoring and Control – Novice), experienced PMs use more complex tools to manage issues. Issues can be categorized in the issue log in a number of ways including:

- Type of issue (i.e., people, process, technology)
- Priority (urgent, high, medium, low)
- Issue age (based on open date)
- Issue severity (based on impact)

These categories will assist the PM to prioritize the issues and help lead the team to resolve them. Using these categories, the PM can create a dashboard to present the status of issues to the stakeholders. Dashboards are a good way to present high-level information to stakeholders so that they can know how the project is progressing without being enmeshed in the details. If the issues and risk logs are maintained in a spreadsheet, dashboards can be generated using Pivot tables. Examples of two ways to present the issues using Excel Pivot tables can be found in Tables 10.1 and 10.2. (More information about Pivot tables can be found at support.office.com.)

Table 10.1   Issues by Summary Pivot Table

Status Summary	Issues		Risks		
Area	Closed	Open	Closed	Open	Total
Content Development	1	4	1	1	7
Credentialed Trainers and Schedule	1	0	0	2	3
Data and Reporting	0	3	0	1	4
Learning Management System	0	1	0	1	2
Logistics and Call Center	0	1	0	1	2
Project Management	1	3	1	0	5
Total	3	12	2	6	23

Table 10.2   Issues by Severity Pivot Table

Status by Severity	Issues		Risks		
Area	High	Medium	High	Medium	Total
Content Development	2	2	0	1	5
Credentialed Trainers and Schedule	0	0	2	0	2
Data and Reporting	2	1	0	1	4
Learning Management System	1	0	0	1	2
Logistics and Call Center	0	1	0	1	2
Project Management	3	0	0	0	3
Total	8	4	2	4	18

Risks can be reported in the same way as issues, although issues often take priority as the risks may not materialize and may lessen as the project progresses. In addition to reporting issues, the dashboard can help focus the PM (and stakeholders) on aging issues. Sometimes, older issues were identified early in the project but were not ready to be addressed at the time they were identified. Other times, however, older issues are the more difficult ones and, if neglected, could derail the project.

Collecting regular status on all issues and risks is necessary to keep the project on track. Responsible team members should update every issue at least weekly, and more frequently in shorter projects or as the project nears go-live. Project team meetings should focus on updates on the highest priority issues and potentially realizable risks. Issues and risks dashboards should be presented to the stakeholders. By using dashboards, the stakeholders can see the overall progress of the issues and risks. Then the status meeting can focus on issues requiring stakeholder involvement. If an issue requires information or decisions from the stakeholders, it should be presented as such. Often posing the issue as a question that needs addressing helps to expedite an answer from the stakeholders. See Table 10.3 for an example of an issue posed as a question.

Including options that are spelled out along with risks, benefits, and potential impacts of each option will help stakeholders to decide the best option. As always, the PM should document decisions, the decision date, and who decided, along with the issue. Once a decision is obtained, the work to resolve the issue can begin.

Table 10.3   Issue Posed as a Question

Issue Requiring Decision from Stakeholders
Issue #: 103
Description: How long after discharge should the orders with a start date and time after discharge be cancelled?
Identifying Project Team: Clinical
Impacted Organization(s): All
Impacted Project Team(s): Clinical Care, Surgery, Lab, Radiology, ED, Pharmacy, Rev Cycle
Decision Required by: 1.1.2020
Has the decision been made: No
Decision Participants: Nurses, Physicians, Medical Records, Laboratory, Radiology, and Billing department
Decision Forum: Have taken this to each team to ask questions and received information about current state from IT.

	Option 1	*Option 2*
Description	Cancel orders at discharge (mimics current state)	Cancel orders 1 hour after discharge (model recommendation)
Plan to Execute Option	• Identify decision and update table	• Identify decision and update table • Potentially modify existing policies/procedures
Risks/Benefits	• Benefit – mimics current state • Risk – If orders are canceled at discharge and discharge is reversed, would need to determine process to 're-instate orders' • Risk – If orders are canceled at discharge and discharge is reversed, orders could be missed when 're-instated'	• Benefit – If orders are canceled at discharge and discharge is reversed within 1 hour, orders will not have been discontinued and can remain as ordered • Risk – does not mimic current state • Risk – Lab/Rad/Pharmacy may have future orders within the 1-hour window and will attempt to draw or carry out the order if they do not see the patient is discharged
Assumptions	• Assumes that new system could 're-instate' the orders as system does today	• Assumes orders remain 'live' for 1 hour after the patient is discharged

Managing Constraints

Projects are assessed in two ways. Overall project metrics include on time, on budget, and within scope. Organizational metrics include how the organization determines if the project accomplished what it set out to do. Both are important to manage and track during the Execution and Control phase. Project metrics are usually measured in tasks completed on time, milestones achieved, number of issues resolved, change requests completed, and budget. Most project management software includes reports that can generate a lot of this information. Using an overall project dashboard can present the information clearly to the stakeholders. See Table 10.4 for an example of an overall project dashboard.

While the triple constraint discussed earlier in this chapter addresses the overall project metrics, other constraints help determine whether the project has actually met the organizations goals (organization metrics). These other constraints include:

■ Quality
■ Risk
■ Resources
■ Sustainability
■ Organization processes
■ Customer satisfaction

Experienced PMs focus on these constraints in addition to the basic triple constraint.

Quality Constraint

While quality is closely related to scope, it often means more to the stakeholders. The quality question asks how closely the outcome metrics match the expectation. Although it is not part of our case study, one project example that many PMs have faced is around order sets. Implementing order sets means many things to many people. Often, the details are not discussed before the start of the project as many have been involved in order to set development and implementation. To many vendors or consultants, there is an assumed number or type of order sets that is included in the project. To quality and risk departments, there is often an assumption that order sets

Table 10.4   Example of an Overall Project Dashboard

Overall Project Status		
Area of Work	*Current Status*	*Trend*
Content Design and Development	R	Y
Data and Reporting	G	G
Credential Trainers	Y	Y
Training Logistics (Call Center)	G	Y
Training Environment	Y	Y
Transition to Post-Live Training	Y	Y
Budget	Y	G

Current and Upcoming Tasks			
Task	*Target*	*Status*	*Comments*
Revise content for delivery	12/28	R	Content is printed, but some are still awaiting final sign-off.
Sign-off of Provider Group 2 content	12/28	Y	Content is printed, but some are still awaiting final sign-off.
Post and send daily report of new Tip Sheet uploads	Ongoing	G	Completed daily

Issues and Risks Status				
Issue/ Risk	*Description*	*Impact*	*Severity (H/M/L)*	*Mitigation*
Issue	Lesson plans require review and final sign-off	Delayed development has forced final sign-off to last minute or even after first delivery.	H	Tracker is updated. Status on reviews and 'readiness' is tracked weekly.
Issue	Final decisions have not been made on the specialty provider tracks	Requests for changes to the curriculum continue to have significant impact to content development, sign-off, scheduling, and registration.	H	CRs have been submitted, and list is finalized for provider tracks. Need to update metadata and LMS.

(Continued)

Table 10.4 Example of an Overall Project Dashboard (Continued)

Issues and Risks Status				
Issue/ Risk	*Description*	*Impact*	*Severity (H/M/L)*	*Mitigation*
Risk	Volumes of backlog work efforts are creating delays and downstream impacts	1. Metadata/curriculum stability and accuracy 2. Timely scheduling and registration 3. Timely and quality content development and printing	H	Escalated to leadership review. Working on priority to complete all the expected tasks and address upcoming impacts as they arrive.

will improve standard of care and reduce errors. To physicians, there is often an assumption that the order sets will save them time. To IT staff, there is often an assumption that all order sets will follow a standard look and feel. Rarely are all of these assumptions included in the initial project plan, and many require additional committees and decision makers outside of the project structure that can completely derail a project's quality.

Using our example of training can demonstrate the quality constraint. Trainers often follow a trainer to trainee ratio to optimize training. If the project estimates for the training event used different trainer to trainee ratios, stakeholders could assume the quality of the training would be negatively impacted. If these assumptions were not addressed during the Planning phase and/or changes agreed upon during the Execution and Control phase, project quality is at risk. As with the triple constraint, a change in quality needs to be balanced with a change in another constraint. In order to maintain expected trainer to trainee rations, for example, additional trainers could be brought on (resources and cost) or the training phase could be extended (timeline). Extending the training phase, however, will likely lead to a delay in the go-live unless other work could be shortened or some training could take place after the go-live. The PM can present the issue to stakeholders as a question using options as discussed earlier. In addition, other quality tools such as Cause and Effect Analysis (also called a Fishbone diagram) could be used to balance quality with the other constraints.

Risk Constraint

Another constraint that can affect overall project success is risk. While we discussed managing risks with impacts and likelihood earlier in this chapter, this risk constraint is about the organization's ability to absorb risk. Some organizations embrace risk and are very innovative. Other organizations only make a change that is likely to be successful. Both can be successful and successfully implemented projects such as opening the new Tower. Managing overall project risk, however, is different for each. For the organization that embraces risk, the PM can be more flexible when identifying risk and expect many risks to be absorbed in the project without major upset. If the hardware was not ready for this organization, they would be likely to manage the opening of the Tower with existing devices or fewer than planned computer carts and would add the additional carts after the opening.

Organizations that avoid risk, however, need to be managed more carefully. Often, projects for these organizations require clear, detailed risk mitigation plans that are very likely to prevent the risk from occurring. At times, this type of organization will even not start or will put a stop to a project with too much risk. An organization that avoids risk would be more likely to delay the Tower opening if enough hardware was not available. Another example to describe this organization's risk avoidance is hardware failure. Unless every device is tested in its new location, there is no guarantee that it will work correctly the first time it's used. Perhaps, information technology (IT) identified a potential 1% failure rate. Mitigation includes IT staff available at go-live to fix any non-functioning device. On paper, a 1% failure rate of the new computer carts does not sound like a huge risk. Having the cart fail or not connect during a critical patient care event such as a code or urgent admission, however, could be considered disastrous to the patient care givers. For a risk intolerant organization, this would constitute a project failure.

Using a tool like the Failure Mode and Effect Analysis (FMEA) can help to prioritize and control overall risk with stakeholders. According to the Institute for Healthcare Improvement (IHI), FMEA is a "systematic, proactive method for evaluating a process to identify where and how it might fail and to assess the relative impact of different failures, in order to identify the parts of the process that are most in need of change" (IHI, 2017). FMEA includes review of the following questions to assess the risk potential:

- What could go wrong? (failure modes)
- Why would the failure happen? (failure causes)
- What would be the consequences of each failure? (failure effects)

The PM can use this tool to create mitigation plans for any potential risk. This will help identify the methods to resolve the potential issue before it occurs. FMEA is intended to correct processes proactively rather than after the risk has become an issue, so it is a good tool for identifying risk mitigation plans. Getting stakeholder buy-in for risk probability and impacts is key for overall project success, no matter the organization's risk tolerance.

Resource Constraint

Another overall project success constraint is resources – human, technical, and financial. While the project metric is about enough resources to complete the project, this organization constraint is about having the *right* resources. Using our training example, the resources are both the trainers and the training materials. Not everyone is able to train others without having some training themselves. In order to create materials for the new Tower, the trainer needs to have an understanding of both current and future processes so they can focus the training appropriately. In addition, the trainer needs to have some skill in creating training materials the way they will be created in this project. If electronic training materials are going to be used, best practice defines that there need to be enough computers for each student to have their own computer for the class. In addition, both paper and electronic materials should work for all types of learners and meet the requirements of the Americans with Disabilities Act (1990) or other regulations for readability such as the National Institutes for Health (NIH) PRISM Readability Toolkit (Ridpath, Greene, & Wiese 2007). Without the right resources to create and deliver the training, this organization constraint may be considered a failure.

Another example of this constraint is the need for a specific IT skill, such as someone who could build new flowsheets in the electronic medical record (EHR) or configure the new computer carts. In addition, the computer carts themselves are a required resource and need to be available in time to test and install them in the new Tower. These resources are essential to the project and must be available at certain times in the timeline. The PM can manage human, technological, and financial resources within the workplan.

Due dates to order, receive, test, and install the computer carts, along with the resources needed at each step, can be added to the workplan. When resources are added to the workplan, the amount of time they will spend on the project is also added. Usually, by default, the resource time will match the task time. While the resources needed to test the computer carts will probably be needed for the duration of the test and install tasks, that is not true for all tasks. For example, if reviewing one policy and procedure (P&P) is scheduled to be completed in 1 month, each of the resources assigned to this task will have a month's worth of time assigned to them. It is unlikely, however, that these resources will spend the entire month reviewing the P&P. Therefore, in addition to the task's duration, the PM needs to estimate the time each resource will spend on each task. This will not affect the task duration but will affect the amount of time each resource is scheduled to spend on the project. For example, if a resource only has 8 hours a week to devote to this project, the total time of all of their tasks per week should not be more than 8 hours. Project management software can greatly help to make sure staff are not over-scheduled within the project.

Sustainability Constraint

The next constraint that the PM needs to manage for organizational success is sustainability – social, environmental, and economic. Social impacts include the basic labor practices, customs, and ethical behavior. Projects need to take into account 'typical' work hours, for example. If no overtime is allowed, any additional work needs to increase the schedule or resource or lead to a reduction in scope. Environmental impacts are less obvious in healthcare projects. One example that could impact this project is a cultural norm about reducing computer 'waste'. Perhaps the organization expects devices to last 5–7 years, and replacing them earlier requires detailed business impacts. If this is the case in our new Tower, the organization may want computer carts that are currently in use in the old Tower to be moved to the new Tower as each unit moves over. This will limit their availability to be tested in the new Tower. The PM needs to balance the risk of available software against the organization's desire to reduce computer 'waste'. Economic impacts include return on investment, business agility, and the economic stability. For the new Tower to be successful, there need to be enough patients to fill the beds, and the downtime due to the move must be minimized.

The Tower building certificate of need would have identified the return on investment of the new beds. It is up to the PM, however, to create an activation plan that minimizes downtime; otherwise, the organization may see this as a project failure.

Organization Processes and Customer Satisfaction Constraints

Organizational process and customer satisfaction are two constraints that go together. Each organization has its own culture, and this often drives how satisfaction is defined. A quote originally attributed to Peter Druker and used by many strategists is that "culture eats strategy." Just as some organizations are risk adverse or tolerant, some organizations expect near perfection in project work. Issues are a normal part of any project, but some organizations are surprised when they see the number of issues identified. They may equate the number of issues with poor work. Experienced PMs expect issues and often encourage teams to document them as a way to reduce surprises later in the project. PMs need to manage to the organization's style in order to drive satisfaction. Using the issues example, in addition to working the issues to resolution, the PM can share the 'typical' number of issues that are generated by projects to help the organization understand that issues are part of any project. In this way, organizations may better understand what is acceptable in a project.

Communication within the organization can also impact customer satisfaction. Projects usually have a sponsor(s), and often this sponsor(s) is responsible for communicating to others outside the project. The sponsor(s) needs to understand how the project is going and be able to explain to the rest of the organization. Again, using the issues example, if the sponsor shares that the project has many issues that are being worked, others in the organization can see this as a problem and assume the project is in trouble. If the sponsor(s) instead shares that the project team is making progress and preventing issues from derailing the project, the organization may have a more positive outlook about the project's ultimate success. The PM needs to understand the nuances of the organization's culture to assist both the sponsor(s) and the team work through the project 'typical' ups and downs. Using the outputs from the Planning phase, RACI (responsible, accountable, consulted, and informed) and stakeholder grids can greatly help the PM manage the work so that the organization (the customer) is ultimately satisfied with the project outcomes.

Testing

In addition to monitoring the work and managing the project issues, risks, and constraints, the PM is often responsible for executing and controlling the testing needed for the project to go live successfully. Test plans include details about what needs to be tested as well as what constitutes a successful test. Most complex projects have numerous test events starting with unit testing each item or device to make sure it works. This work is often assumed as part of the 'build' task and does not always get a separate step in the project. Depending on what is being 'built', integration, system, regression, and acceptance testing could also be part of the overall test plan. Using our hardware example, computer carts and handhelds should be tested to make sure they are able to access all the applications required and connect from the actual ports on the new Tower. Using our electronic medical record (EMR) revisions, patient rooms should be the same in all of the systems, so a test could include moving a patient in the EMR and assessing whether that change displays correctly in the laboratory, pharmacy, billing systems, etc. During the testing, issues as well as success metrics should be gathered and reported. In addition, if the test does not go smoothly, the PM will need to work on mitigation plans. For example, the issues could be resolved and then tested again. These steps continue until all of the testing is complete. Often, a project includes a go/no-go decision at the end of the testing work. The PM is responsible for gathering the results of the build and testing and presenting that information to the sponsor(s) and stakeholders to determine if the project should continue to the next steps of training and activation.

Activation

Activation usually includes a period of time before and after the actual go-live and includes readying the command center, creating the go-live support schedule, any needed training for resources who will support the go-live, and how issues will be managed during the go-live and post go-live events. The PM's role is to execute the activation plan and change from 'build' to go-live support. Communication and project control should be the main focus during this part of the project. Status of the project can be reported daily or more often in complex projects and should include the dashboards and other communication tools used throughout the project.

All the stakeholders should be involved in the activation status meetings so that they can provide accurate information to their departments. Often, this is the first time many staff will see the new policies, use the new hardware, and care for patients in the new Tower. They will have questions about how their work differs and will most likely identify issues with the policies, hardware, and applications that have been modified for the new Tower. The PM's main focus is to prioritize and manage identified issues. Issues preventing care or billing should be the highest priority. Other issues may be able to wait until after the go-live event to determine to best way to resolve them. Making a lot of changes to hardware or software during a go-live often leads to errors and additional changes later. It is often the PM who maintains the control over which issues get resolved during the go-live, and which wait until they can be further analyzed for impact. Once the go-live is complete, it is time to close out the project.

Chapter 11

Closing – Expert

This chapter will cover the Closing process group, building on Chapter 7 (Closing – Novice), for experienced project managers (PMs) and organizations with formal and mature processes. In these instances, there are defined procedures in place for the expected activities that occur during this process group. Project closing begins when all deliverables (product, service, or result) have been produced and accepted. This typically occurs at the project go-live or activation.

Finalize Project Documentation

As discussed in Chapter 7 (Closing – Novice), all documentation should be finalized and archived. These become historical information for future projects. By reviewing previous project documentation, the PM can better plan their projects. The historical documentation provides details such as what risks or issues can arise, how they were managed, what tasks were required to complete specific activities, and even how to manage stakeholders. There are many activities that should be completed during closing.

The PM should review all project documentation and ensure they are up to date and complete. The review of the issues list would be to verify all are resolved and all documentation is complete, including the actual resolution. If any issues remain, they should be evaluated with the sponsor(s) for a decision if they need to be resolved prior to the project officially closing or if they can be assigned and resolved outside of the project. If they will be resolved outside the project, they must be assigned to someone who will be responsible for ensuring the resolution and completing the documentation on how it was resolved. The decision and assignment should be documented as an update to the issue and included in the project's completion document.

Project and Organizational Metrics

The PM should ensure the project metrics are compiled and reported according to the metrics plan, or project scope. Chapter 9 (Planning – Expert) outlines the difference between project and organizational metrics or measures of success. All project metrics should be finalized and reported in a specific report or as part of the project completion document. While most organizational metrics are completed outside of the project, there may be an expectation that baseline measurements will be collected during the project. If this is the case, the PM should ensure they have been completed at the expected timeframe.

All other documents that are part of the project management plan (PMP) should be finalized. Each document should be updated to reflect what actually occurred during the project. Depending on the organization's defined process, this may be done by producing a new version of the document or entering updates in a new section at the end so the original content is not changed. These documents include the workplan, stakeholder management plan, test plan, activation plan, and any other PMP documents included in the project.

Support

The project team typically provides support during the initial period of time after the activation. The duration of this activity would depend on the amount of change to the users. For most projects this may last 1–3 weeks. During this time there may be super users on the unit to provide any just-in-time training and be a conduit for reporting actual issues. The project team works to resolve any initial issues prior to the official transition to ongoing support. Prior to the transition, the project team should meet with the support team for knowledge transfer related to the final deliverables and changes included in the project. The transfer also should include providing the users with direction on how to obtain support once the project team is no longer involved. Communication related to requests for change should also be included, which may include standard configuration and release management procedures, or the availability of ongoing training for current and new staff.

Contract Closeout

If there are any contracts in place for the project, they should be properly closed out. The contracts could be for products, software, or staff augmentation. During project closing, the PM should work with the contracting staff to verify all work defined in the contract has been completed to the organization's satisfaction. The PM would also provide any information requested related to performance, metrics, and activities that are needed to officially close out the contract. This activity is repeated for each contract, if there are multiple.

Lessons Learned

The process of collecting and documenting lessons learned has traditionally been completed during project closing. This is a retrospective review of what happened during the project focused on identifying lessons that will help improve processes for future projects. With this in mind, it is important to have these documented as lessons, what should be done in the future to improve processes. These may be voiced as something that went wrong or did not happen as well as expected. They may also be voiced as something that went well and really provided a benefit to the project. Either way, they are lessons to continuously improve the project processes.

There are some PMs who work to gather the lessons throughout the project, not just at the end. This is a valuable practice for lengthy projects where some lessons may not be remembered. If a project will be implemented in phases, a lesson learned meeting should occur after each phase. This provides an opportunity to improve for each subsequent phase.

As mentioned above, these are lessons and should be documented as such. They should state what should be done and why. The why is important to provide context and allow the PM to determine if it is appropriate for their project. They should be written as if they will be read a year from now, but someone who knows nothing about the project. Below are a few examples of comments from team members and examples of how the lessons could be documented. Once these are collected, they should be documented according to the organization's procedure. Some use documents while others may have a database that provide a search functionality.

Comment – Our team had five other projects that we had to manage while trying to meet deadlines for this one, and it was not clear which project took priority.

Lesson – Before beginning a new project, the PM should be aware of its priority related to others in progress. This will set clear expectations if there are conflicts with resources and schedules.

Comment – The activation checklist helped keep everyone on task, but it would have been nice to have a few rehearsals to make sure we understood the full implementation plan.

Lesson – An activation rehearsal should be scheduled for each project to verify the right tasks are in the right order and with the right durations. This also helps allow the team to practice if any tasks are new (new system). If the rehearsal goes wrong, repeat.

Comment – The night shift felt neglected and did not have full support of super users and on-site support staff, the way the day shift did. Thus, we had to learn the system on our own.

Lesson – When preparing for the activation, ensure each shift is covered for end user support. This ensures all staff has access to resources as they begin to use the system.

The final activities during project closing are to celebrate all the hard work and a successful finish. The resources can then be released from the project to allow for new assignments. The PM should then follow the defined procedure for administrative closeout. This is an internal activity to actually mark the project as done. This may include a final status report, updating a project list with the completion date or communication to the stakeholders regarding the completion status.

This chapter reviewed a number of the activities that occur during project closing. Ensuring complete documentation that is archived provides historical and reference materials for future projects. Lessons learned documentation allows for continuous improvement of project processes. The completion document provides a gap analysis of what was planned compared to what actually occurred and when signed by the sponsor(s) and indicates their acceptance and approval to end the project. Always remember to celebrate the team's success.

References and Additional Reading

Books

A Guide to the Project Management Body of Knowledge (PMBOK Guide),
     6th ed. Newtown Square, PA: Project Management Institute, 2017. ISBN:
     978-1628251845.
Finnell, John T., & Dixon, Brian E., eds. *Clinical Informatics Study Guide*.
     New York, NY: Springer, 2016. ISBN: 978-3319227528.
Harris, J., Roussel, L., Dearman, C., & Thomas, P. *Project Planning and
     Management: A Guide for Nurses and Interprofessional Teams*, 3rd ed.
     Burlington, MA: Jones & Bartlett Learning, 2020. ISBN: 978-1284089837.
Harvard Business Review (HBR). *HBR Guide to Project Management*. Boston, MA:
     Harvard Business Review Press, 2012. ISBN: 978-1422187296.
Houston, Susan M. *The Project Managers Guide to Health Information Technology
     Implementation*, 2nd ed. Boca Raton, FL: CRC Press, Taylor & Francis Group,
     2018. ISBN: 978-1138626232.
McCormick, K., Gugerty, B., & Mattison, J. (2017). *Healthcare IT Exam Guide
     for CHTS and CAHIMS Certifications*, 2nd ed. New York, NY: McGraw-Hill
     Education, 2017. ISBN: 978-1259836978.
Note, Margot. *Project Management for Information Professionals*. Waltham, MA:
     Chandos Publishing, 2015. ISBN: 978-0081001271.
Schwalbe, K., & Furlong, D. *Healthcare Project Management*, 2nd ed. Minneapolis,
     MN: Schwalbe Publishing, 2017. ISBN: 978-1976573279.
Schwalbe, Kathy. *Information Technology Project Management*, 9th ed. Australia:
     Thomson/Course Technology, 2014. ISBN: 978-1337101356.
Senstack, Patricia, & Boicey, Charles., eds. *Mastering Informatics: A Healthcare
     Handbook for Success*. Indianapolis, IN: Sigma Theta Tau International, 2015.
     ISBN: 978-1938835667.

Articles

2017 Hospital Construction Survey: Construction Projects. Retrieved June 21, 2019 from https://www.hfmmagazine.com/ articles/2749-hospital-construction-survey-construction-projects

3 Reasons Why Projects Fail and How to Avoid Them (2019). Retrieved December 31, 2019 from https://www.villanovau.com/resources/ project-management/why-projects-fail-how-to-avoid-them/

Benner, P. (1982). From Novice to Expert. *American Journal of Nursing*, 82(3): 402–407.

Benner, Patricia. (2004). Using the Dreyfus Model of Skill Acquisition to Describe and Interpret Skill Acquisition and Clinical Judgment in Nursing Practice and Education. *Bulletin of Science, Technology & Society*, 24(3): 188–99. https://doi. org/10.1177/0270467604265061.

Benz, M. (2018). *10 Project Constraints That Endanger Your Project's Success.* Retrieved December 31, 2019 from https://www.projectmanager.com/ blog/10-project-constraints-that-endanger-your-projects-success

Cooke-Davies, Terence J., & Andrew Arzymanow. (2003). The Maturity of Project Management in Different Industries: An Investigation into Variations between Project Management Models. *International Journal of Project Management*, May 28–31, 2002, 21(6): 471–78. https://doi.org/10.1016/S0263-7863(02)00084-4

IHI. (2017). Failure Modes and Effects Analysis (FMEA) Tool. Retrieved March 20, 2020 from http://www.ihi.org/resources/Pages/Tools/ FailureModesandEffectsAnalysisTool.aspx

Feedback Drives Health Facility Design Processes. Retrieved June 21, 2019 from https://www.hfmmagazine.com/articles/2665-feedback-drives-health-facility-design-processes

Kantor, B. (2018). *The RACI matrix: Your Blueprint for Project Success.* Retrieved January 5 from https://www.cio.com/article/2395825/project-management-how-to-design-a-successful-raci-project-plan.html

Kinser, J. (2008). The Top 10 Laws of Project Management. Paper presented at *PMI® Global Congress 2008—North America*, Denver, CO. Newtown Square, PA: Project Management Institute.

Larson, R., & Larson, E. (2009). Top Five Causes of Scope Creep … and What to Do About Them. Paper presented at *PMI® Global Congress 2009—North America*, Orlando, FL. Newtown Square, PA: Project Management Institute.

Leonard, M., Bonacum, D., & Graham, S. (2017). *Institute for Healthcare Improvement SBAR: Situation-Background-Assessment-Recommendation.* Retrieved September 26, 2019 from https://www.lsqin.org/wp-content/ uploads/2017/08/SBARTechniqueforCommunication.pdf

Overview of PivotTables and Pivot Charts, Retrieved December 31, 2019 from https://support.office.com/en-us/article/overview-of-pivottables-and-pivotcharts-527c8fa3-02c0-445a-a2db-7794676bce96

Parkinson, C. (1957). *Parkinson's Laws and Other Studies in Administration.* Boston, MA: Houghton Mifflin Company. Retrieved October 13, 2019 from http://sas2.elte.hu/tg/ptorv/Parkinson-s-Law.pdf.

Ridpath JR, Greene SM, & Wiese CJ; *PRISM Readability Toolkit*, 3rd ed. Seattle, WA: Group Health Research Institute, 2007. Retrieved December 31, 2019 from https://www.nhlbi.nih.gov/files/docs/ghchs_readability_toolkit.pdf

Rhoads, J. (1977). Overwork. *JAMA*, 237(24): 2615–2618. doi:10.1001/jama.1977.03270510037018

Smooth Operator: How to Set up a Hospital Ward. Retrieved June 21, 2019 from https://www.nursingtimes.net/roles/nurse-managers/smooth-operator-how-to-set-up-a-hospital-ward/5026844.article

The Standish Group (Standish). (1994). *The Chaos Report.* Retrieved August 6, 2019 from https://www.projectsmart.co.uk/white-papers/chaos-report.pdf.

The Standish Group (Standish). (2015). *The Chaos Report.* Retrieved August 6, 2019 from https://www.standishgroup.com/sample_research_files/CHAOSReport2015-Final.pdf.

Websites

CIO Magazine. www.cio.com

Healthcare Information Management Systems Society (HIMSS). www.himss.org

Project Management Institute (PMI). www.pmi.org

Index

Note: *Italicized* page numbers refer to figures, **bold** page numbers refer to tables.